JUMP Math 1.2

Book 1 Part 2 of 2

Contents

jump math™

MULTIPLYING POTENTIAL.

JUMP Math
One Yonge Street, Suite 1014
Toronto, Ontario M5E 1E5
Canada
www.jumpmath.org

Writer: Dr. John Mighton
Consultants: Dr. Heather Betel, Dr. Anna Klebanov, Dr. Sohrab Rahbar, Dr. Sindi Sabourin
Editors: Megan Burns, Liane Tsui, Natalie Francis, Katie Baldwin, Jackie Dulson, Janice Dyer, Neomi Majmudar, Una Malcolm, Jodi Rauch
Layout and Illustrations: Linh Lam, Fely Guinasao-Fernandes, Sawyer Paul, Marijke Friesen, Pam Lostracco
Cover Design: Blakeley Words+Pictures
Cover Photograph: © iStockphoto.com/George Bailey

ISBN 978-1-927457-33-7

Fourth printing June 2022

Printed and bound in Canada

A note to educators, parents, and everyone who believes that numeracy is as important as literacy for a fully functioning society

Welcome to JUMP Math

Entering the world of JUMP Math means believing that every child has the capacity to be fully numerate and to love math. Founder and mathematician John Mighton has used this premise to develop his innovative teaching method. The resulting resources isolate and describe concepts so clearly and incrementally that everyone can understand them.

JUMP Math is comprised of Teacher Resources, Digital Lesson Slides, student Assessment & Practice Books, assessment tools, outreach programs, and professional development. All of this is presented on the JUMP Math website: **www.jumpmath.org**.

The Teacher Resource is available on the website for free use. Read the introduction to the Teacher Resource before you begin using these materials. This will ensure that you understand both the philosophy and the methodology of JUMP Math. The Assessment & Practice Books are designed for use by students, with adult guidance. Each student will have unique needs and it is important to provide the student with the appropriate support and encouragement as they work through the material.

Allow students to discover the concepts by themselves as much as possible. Mathematical discoveries can be made in small, incremental steps. The discovery of a new step is like untangling the parts of a puzzle. It is exciting and rewarding.

Students will need to answer the questions marked with a in a notebook. Grid paper notebooks should always be on hand for answering extra questions or when additional room for calculation is needed.

Contents

PART 1

Unit 1: Operations and Algebraic Thinking: Counting and Comparing

Unit 2: Number and Operations in Base Ten: Introduction to Greater Than and Less Than

Unit 3: Operations and Algebraic Thinking: Introduction to Addition

PART 2

Unit 1: Number and Operations in Base Ten: Understanding Place Value

Unit 2: Number and Operations in Base Ten: Using Place Value to Add and Subtract

Unit 3: Operations and Algebraic Thinking: Problem Solving with Pictures, Models, and Equations

Unit 4: Measurement and Data: Telling and Writing Time

Unit 5: Geometry: Reasoning with Shapes

Unit 6: Measurement and Data: Representing and Interpreting Data

NBT1-11 Counting to 100

☐ Shade the numbers as you count them.

1. 20 to 29

11	12	13	14	15	16	17	18	19	20
21	22	23	24	25	26	27	28	29	30

2. 30 to 39

21	22	23	24	25	26	27	28	29	30
31	32	33	34	35	36	37	38	39	40

3. 40 to 49

31	32	33	34	35	36	37	38	39	40
41	42	43	44	45	46	47	48	49	50

4. 50 to 59

41	42	43	44	45	46	47	48	49	50
51	52	53	54	55	56	57	58	59	60

61	62	63	64	65	66	67	68	69	70
71	72	73	74	75	76	77	78	79	80
81	82	83	84	85	86	87	88	89	90
91	92	93	94	95	96	97	98	99	100

☐ Use the chart to fill in the blanks.

5.

61, 62, 6___, 6___, 6___, 66

71, 7___, 7___, 7___, 7___, 76

81, _____, _____, _____, _____, 86

91, _____, _____, _____, 95, _____

☐ Fill in the blanks.

6.

54, 55, _____, _____, _____, _____

91, 92, _____, _____, _____, _____

80, 81, _____, _____, _____, _____

7. Kim counts from 30 to 37.
She writes 30, 31, 32, 33, 84, 35, 36, 37.
Explain her mistake.

Count all the letters.

☐ Write the new total after each word.

☐ Check with a partner. Correct any mistakes.

8.

6	8	17
Autumn	is	sometimes
called	fall	because
		47
the	leaves	fall
from	many	trees.
63		
The	leaves	can
turn	red,	brown,
yellow,	or	orange.

NBT1-12 Numbers Before and After

☐ Shade the numbers.

☐ Circle the numbers as you shade them.

1.

(1) (5) 8 9 10 11 12 14 15 18 20

21 23 25 28 29 30 31 35 38 40

1	2	3	4	5	6	7	8	9	10
11	12	13	14	15	16	17	18	19	20
21	22	23	24	25	26	27	28	29	30
31	32	33	34	35	36	37	38	39	40

2.

72 73 74 77 79 83

87 88 89 93 97 99

71	72	73	74	75	76	77	78	79	80
81	82	83	84	85	86	87	88	89	90
91	92	93	94	95	96	97	98	99	100

☐ Write the letters you made.

3.

_____ _____ _____ _____

☐ Circle the number on the chart.
☐ Write what comes **after**.
☐ Write what comes **before**.

1	2	3	4	5	6	7	8	9	10
11	12	13	14	15	16	17	18	19	20
21	22	23	24	(25)	26	27	28	29	30
31	32	33	34	35	36	37	38	39	40
41	42	43	44	45	46	47	48	49	50

4. 24 25 26

5. _____ 8 _____

6. _____ 37 _____

7. _____ 21 _____

8. _____ 40 _____

9. _____ 19 _____

10. _____ 44 _____

11. _____ 11 _____

☐ Fill in the missing numbers.

12.

33	34	35		37
43	44	45	46	47
53	54	55	56	57

13.

11	12	13	14
21	22	23	24
31		33	34

14.

21	22	23
31	32	33
41	42	
51	52	53

15.

64	65	66
	75	76
84	85	86
94	95	

16.

48	49
58	59
	69
78	79

17. BONUS

16	17	18	19	20
26	27	28	29	30
36		38	39	40
46	47	48	49	
56	57	58	59	60

18. BONUS

32		34	35	
42	43	44		46
	53	54	55	56
62	63	64	65	
72	73		75	76

NBT1-13 Tens

1	2	3	4	5	6	7	8	9	10
11	12	13	14	15	16	17	18	19	20
21	22	23	24	25	26	27	28	29	30

20 is 2 tens blocks

☐ Use tens blocks to find the answer.

1	2	3	4	5	6	7	8	9	10
11	12	13	14	15	16	17	18	19	20
21	22	23	24	25	26	27	28	29	30
31	32	33	34	35	36	37	38	39	40
41	42	43	44	45	46	47	48	49	50

1. 30 is _____ tens blocks.

2. 40 is _____ tens blocks.

3. 50 is _____ tens blocks.

4. 10 is _____ tens block.

☐ Now answer without using blocks.

5. 60 is _____ tens blocks.

6. 70 is _____ tens blocks.

☐ Fill in the blanks.

7.

__2__ tens blocks = __20__ ones blocks

8.

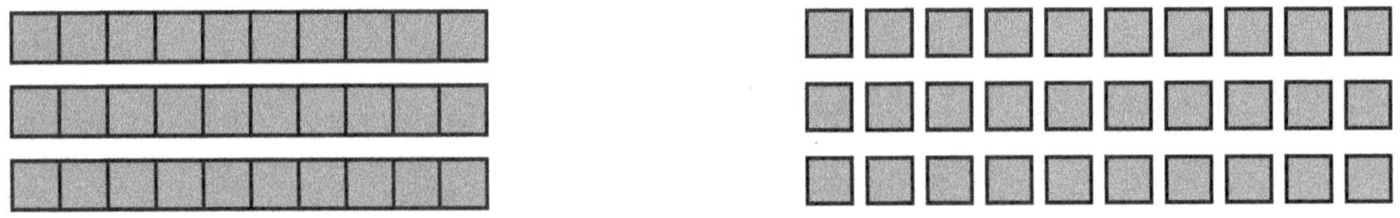

_____ tens blocks = _____ ones blocks

9.

_____ tens blocks = _____ ones blocks

10. 5 tens blocks = _____ ones blocks

11. 6 tens blocks = _____ ones blocks

12. 7 tens blocks = _____ ones blocks

☐ Fill in the blanks.

13.

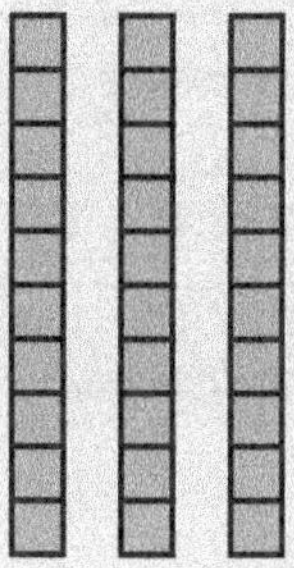

3 tens = 30 ones

14.

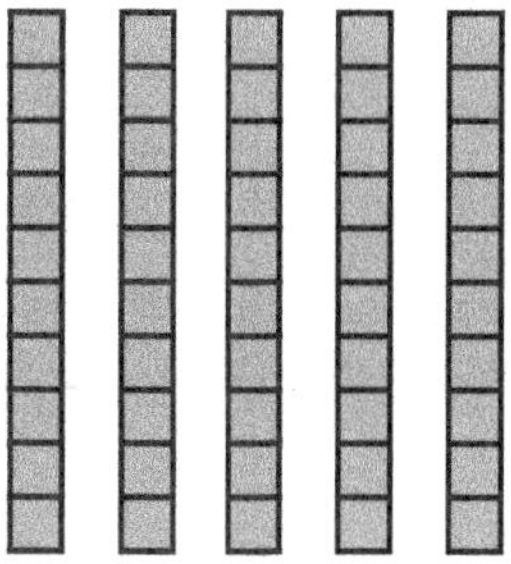

_____ tens = _____ ones

15.

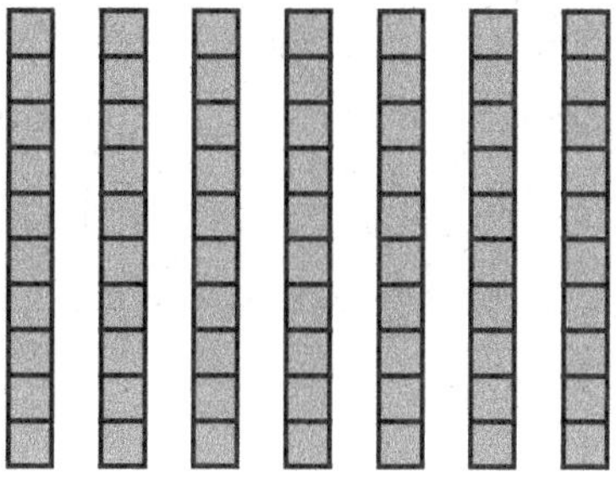

_____ tens = _____ ones

16.

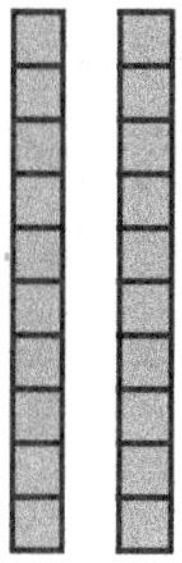

_____ tens = _____ ones

17.

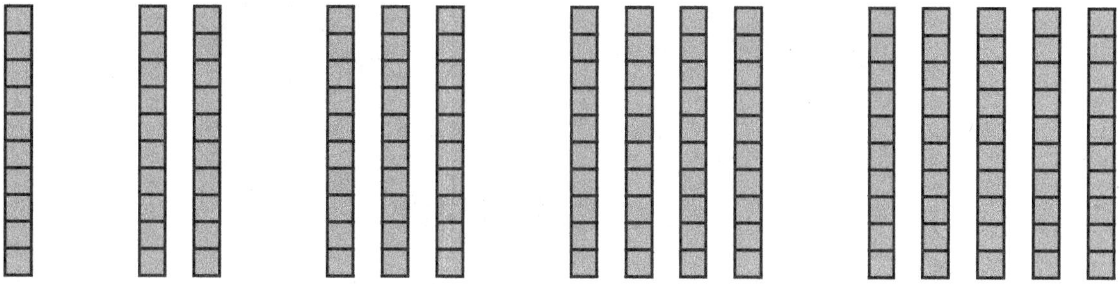

10 20 _____ _____ _____

☐ Count by tens.

18. 10, 20, _____, _____, _____

19. 40, 50, _____, _____, _____

NBT1-14 Counting Groups of 10

☐ Count how many. Use groups of 10.

1.

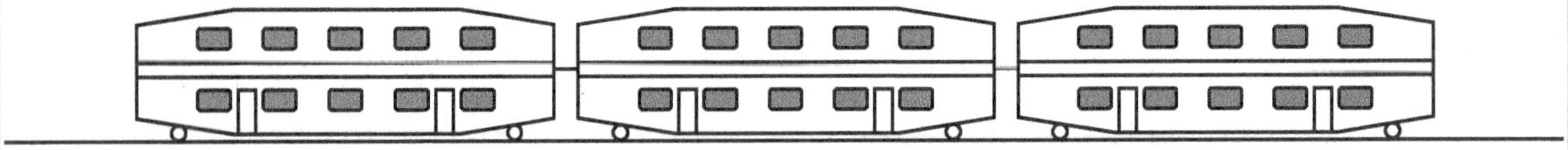

_____ windows

2.

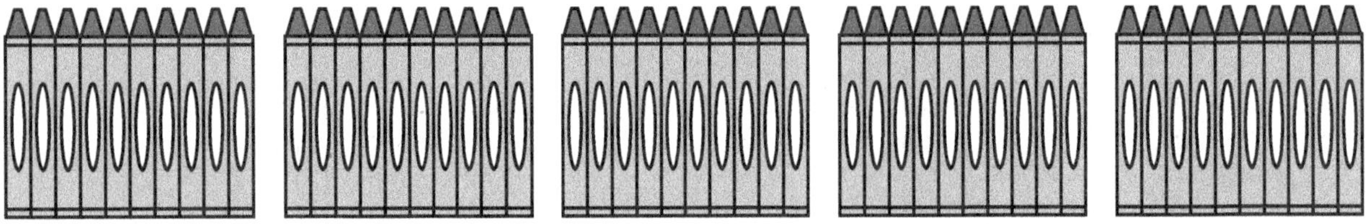

_____ crayons

3.

_____ apples

4.

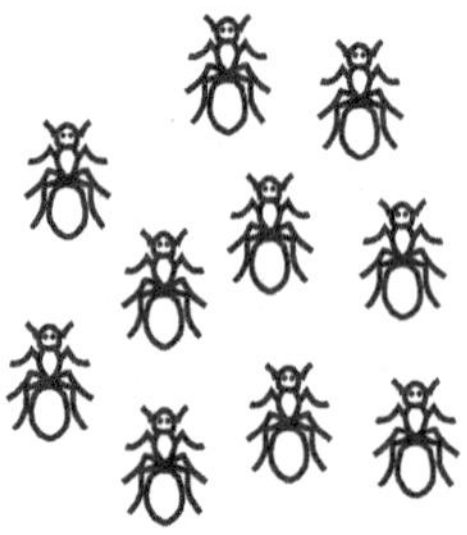

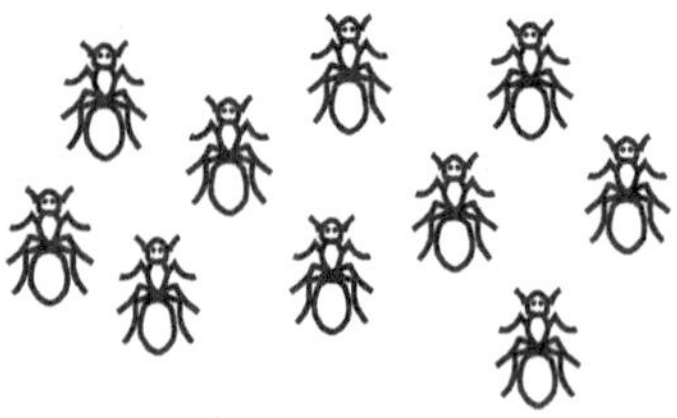

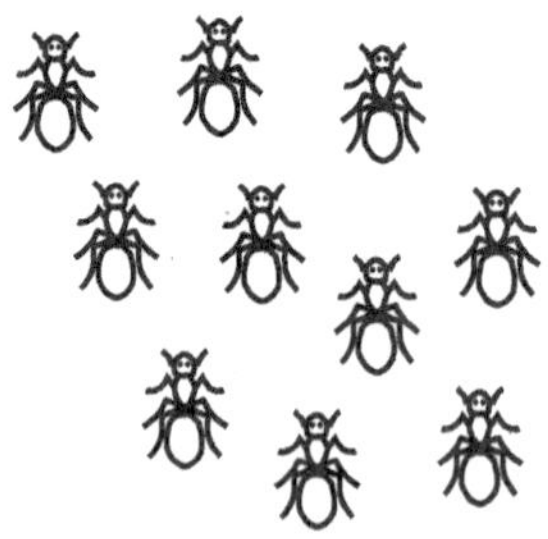

_____ ants

☐ Circle groups of 10.

☐ Write the number of ones blocks.

5.

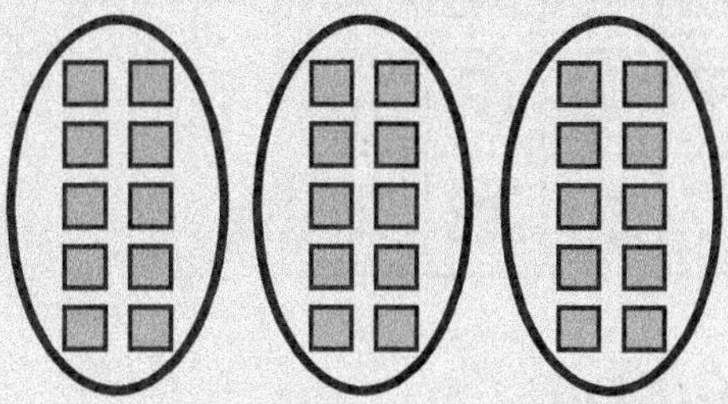

30 ones blocks

6.

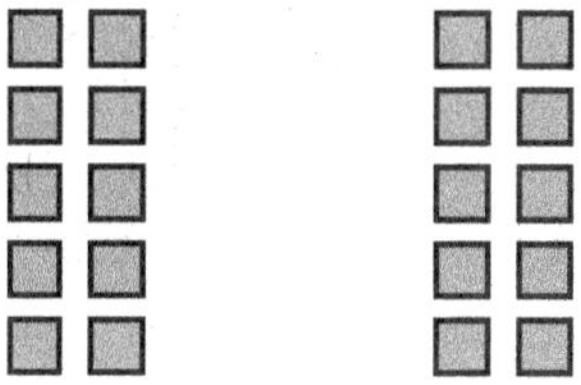

_____ ones blocks

7.

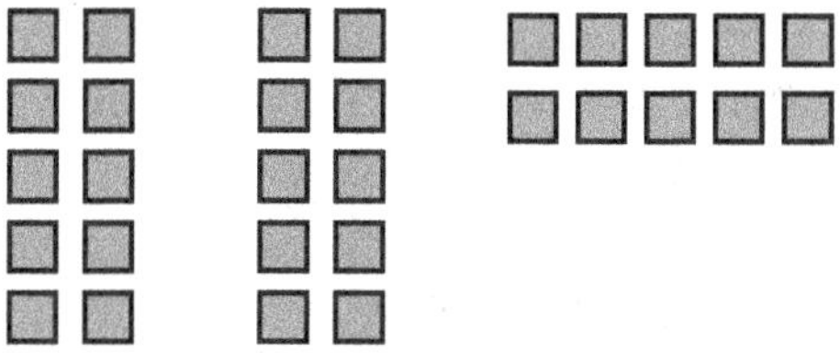

_____ ones blocks

8.

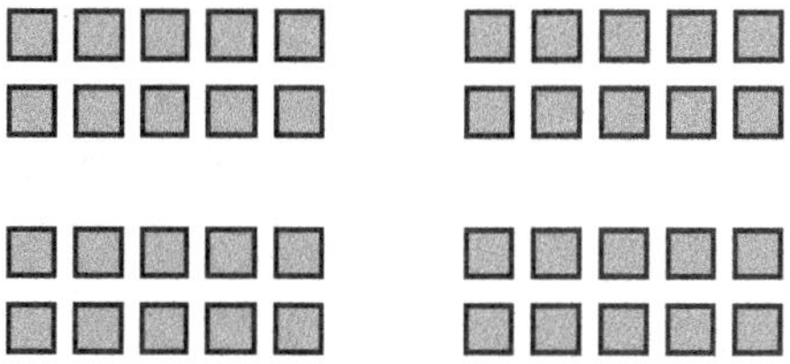

_____ ones blocks

9.

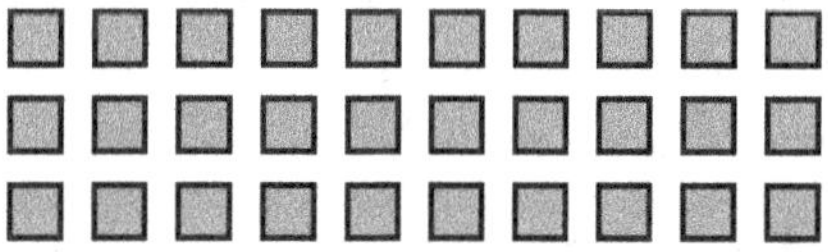

_____ ones blocks

10.

_____ ones blocks

11. Jon says the picture shows 20 dots. Explain his mistake.

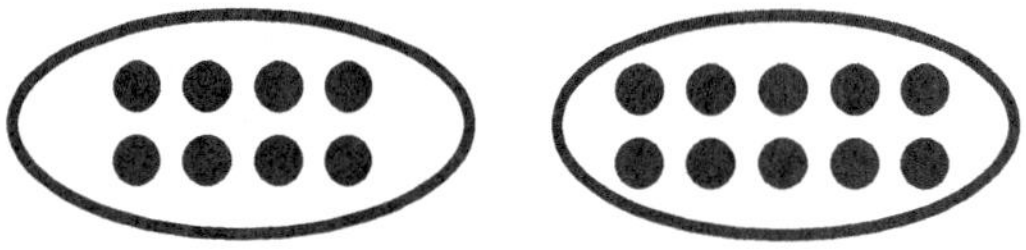

12. Sharon puts 10 pennies in her piggy bank each day. How many pennies does she have after 4 days?

NBT1-15 Tens and Ones Digits

1	2	3	4	5	6	7	8	9	10
11	12	13	14	15	16	17	18	19	20
21	22	23	24	25	26	27	28	29	30

24 is 2 tens and 4 ones.

☐ Use the chart to fill in the blanks.

1	2	3	4	5	6	7	8	9	10
11	12	13	14	15	16	17	18	19	20
21	22	23	24	25	26	27	28	29	30
31	32	33	34	35	36	37	38	39	40

1. 28 is _____ tens and _____ ones.

2. 35 is _____ tens and _____ ones.

3. 27 is _____ tens and _____ ones.

4. 30 is _____ tens and _____ ones.

- [ ] Fill in the chart.
- [ ] Write the number shown.

5.

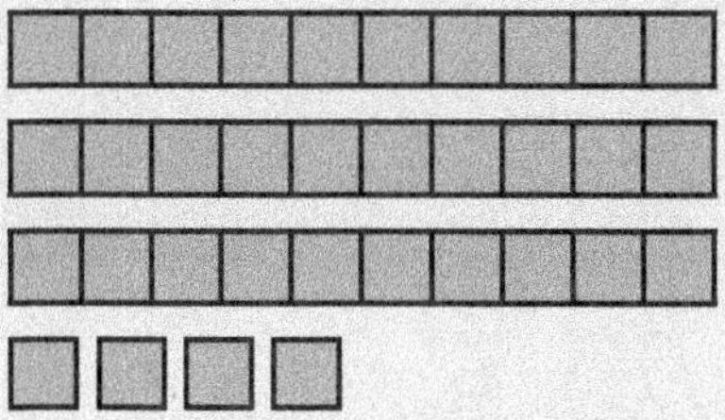

Tens	Ones
3	4

number = 34

6.

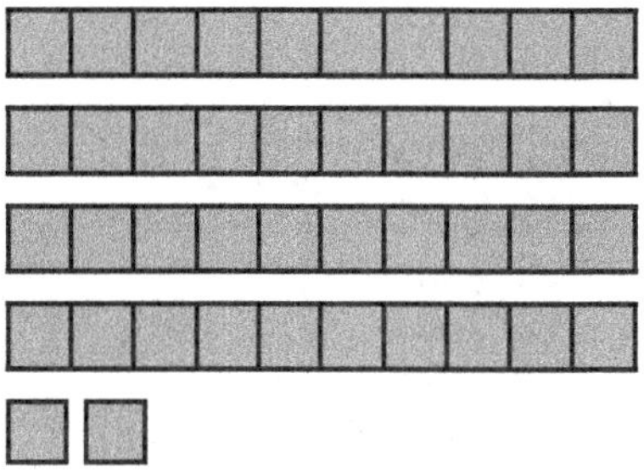

Tens	Ones

number = _____

7.

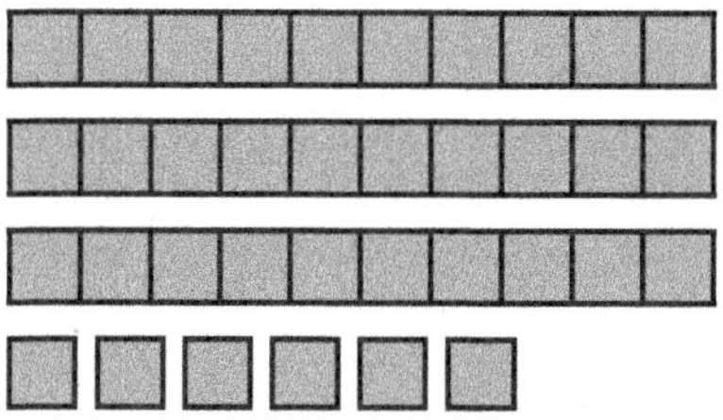

Tens	Ones

number = _____

8.

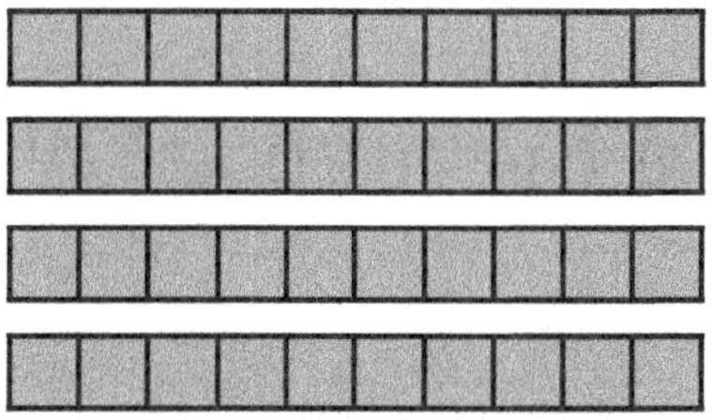

Tens	Ones

number = _____

- [ ] Show each number using blocks.

9. 50 43 37 19 32

31

tens digit ← → ones digit

☐ Circle the digit.

10. (3) 2 — tens digit

11. 1 (7) — ones digit

12. 8 1 — tens digit

13. 7 5 — tens digit

14. 6 3 — ones digit

15. 5 2 — ones digit

16. 9 8 — tens digit

17. 6 6 — ones digit

18. 2 3 — tens digit

☐ Write the number of ones or tens.

19.

The 3 in 37 means *3 tens*.

The 6 in 46 means *6 ones*.

The 2 in 92 means ______________.

The 3 in 35 means ______________.

20. How many ones are in 2 tens blocks?

21. Don uses 3 tens blocks to show a number.
What number does he make?

NBTI-16 Tens and Ones (Advanced)

☐ Add a ones block.

☐ Write the new number.

1.

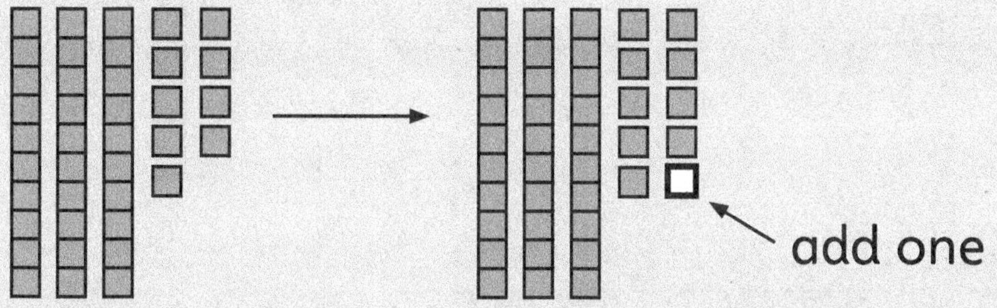

39 → 40

2.

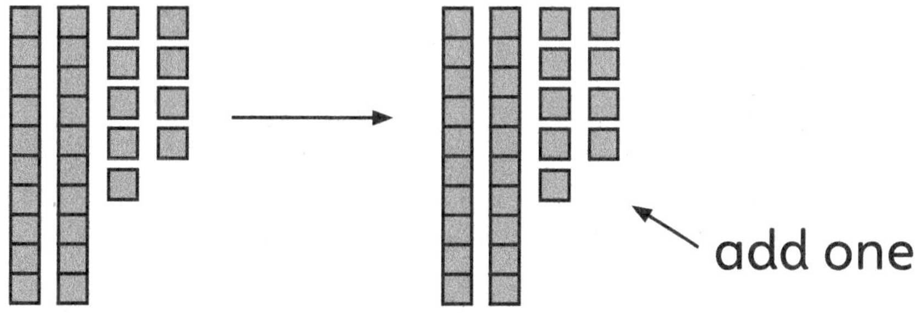

29 → ______

☐ Fill in the missing number.

3. 39, ______, 41

4. 59, ______, 61

5. 29, ______, 31

6. 49, ______, 51

7. 89, ______, 91

8. 69, ______, 71

☐ How many ones in all?

9.

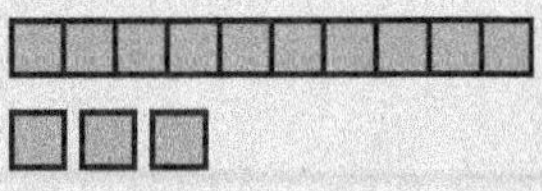

_1_ ten + _3_ ones = _13_

10.

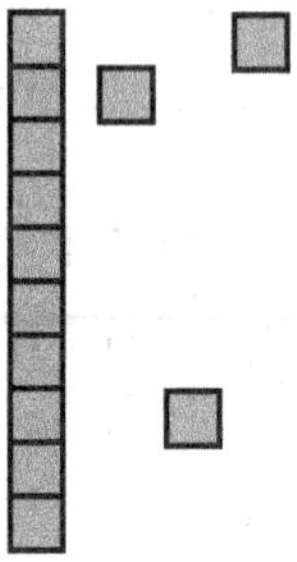

___ ten + ___ ones = ____

11.

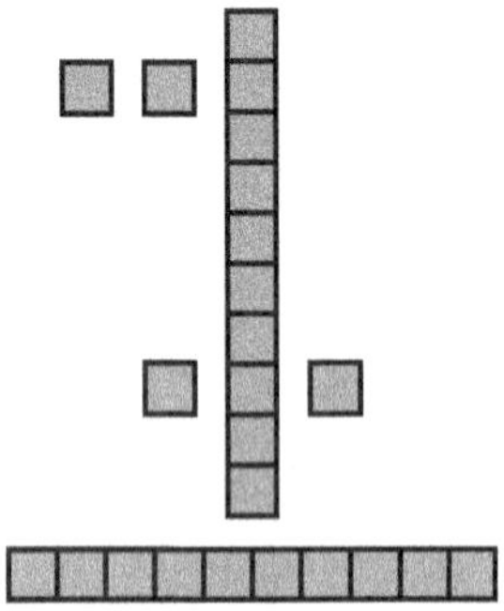

___ tens + ___ ones = ____

12.

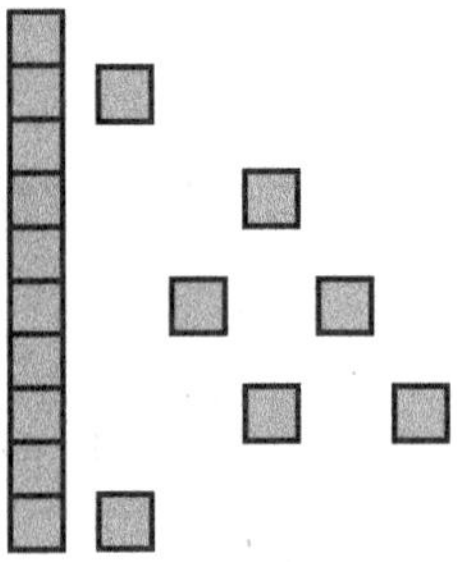

___ ten + ___ ones = ____

13.

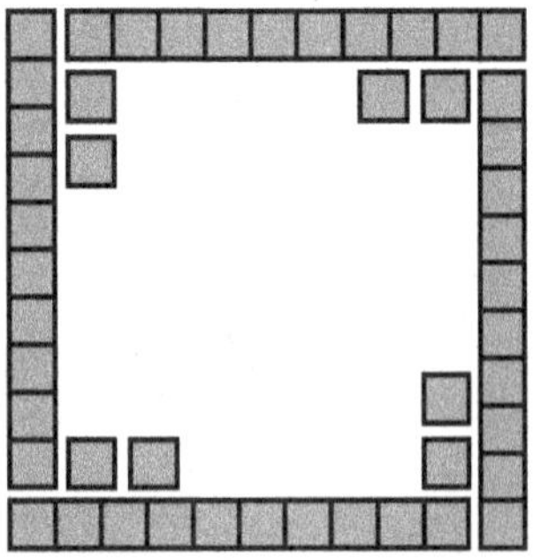

___ tens + ___ ones = ____

14.

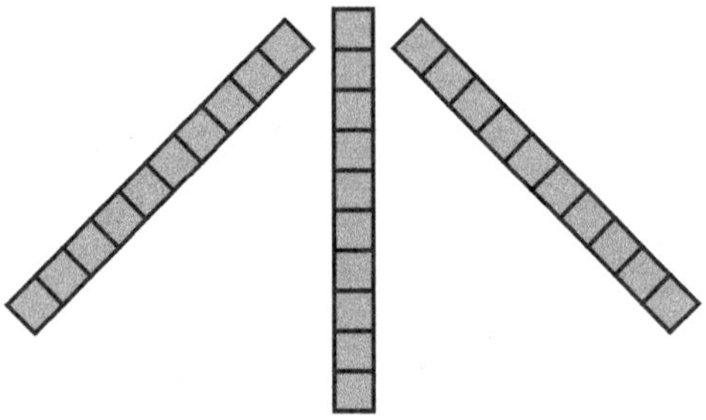

___ tens + ___ ones = ____

You can draw a line | for 

You can draw a dot • for ▫

☐ Draw | and • to show the number.

15.

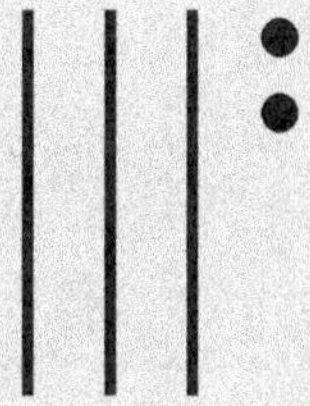

32

16.

24

17.

43

☐ Write the number.

18.

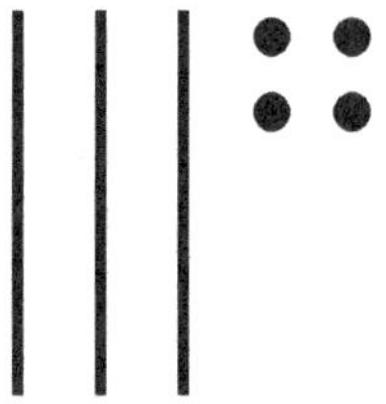

_____

19.

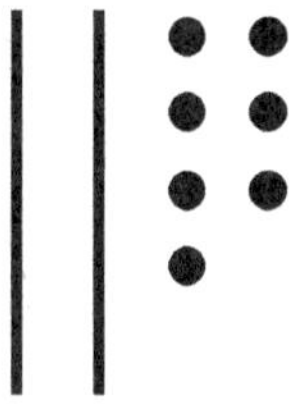

_____

20.

_____

21. BONUS

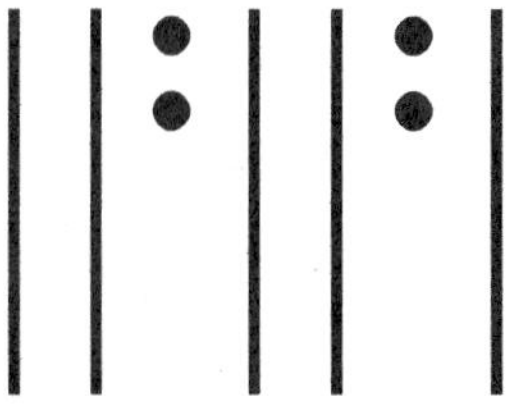

_____

22. BONUS

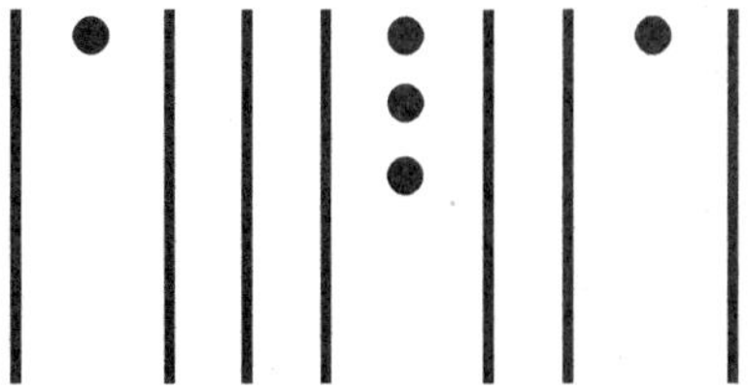

_____

NBT1-17 Comparing Numbers Up to 50

- [ ] Use base ten blocks to show each number.
- [ ] Circle the **larger** number.

1. (23) or 15

2. 26 or 30

3. 40 or 45

4. 21 or 17

5. 7 or 11

6. 30 or 28

7. 26 or 29

8. 32 or 36

9. 35 or 41

10. 42 or 24

11. 13 or 31

12. 64 or 46

13. 71 or 69

☐ Can you shade 25? Write **yes** or **no**.

☐ Write the number shown.

☐ Is the number **greater** or **less** than 25?

14.

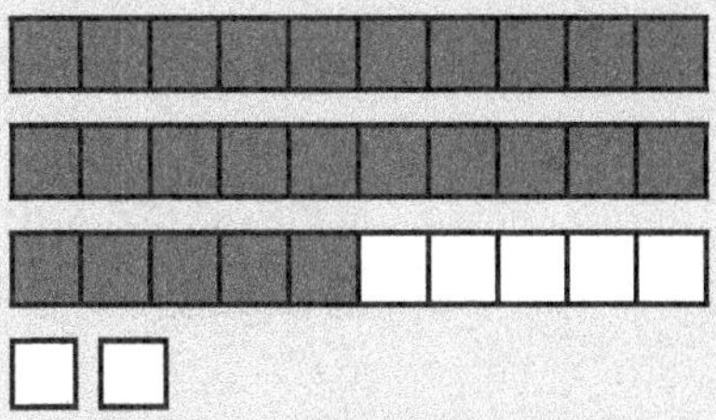

yes

32 is *greater* than 25.

15.

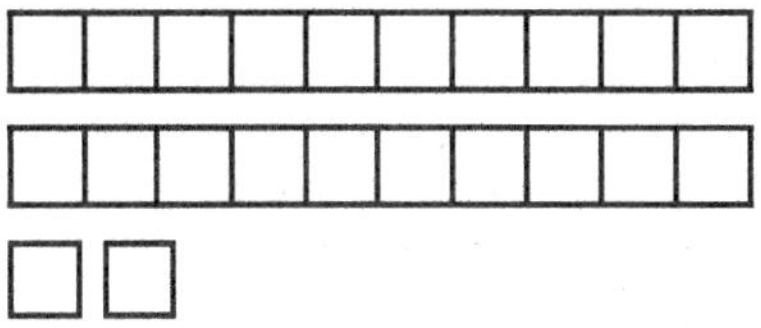

no

____ is __________ than 25.

16.

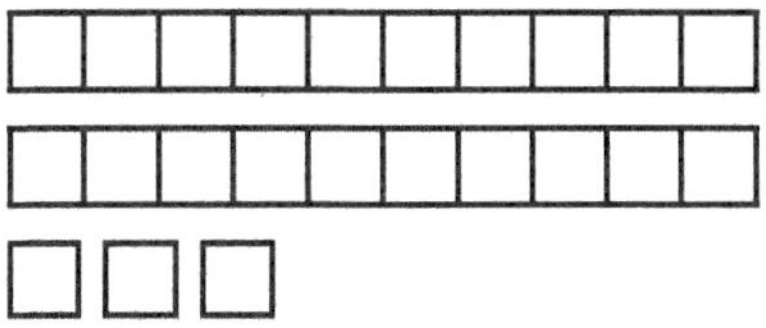

______

____ is __________ than 25.

17.

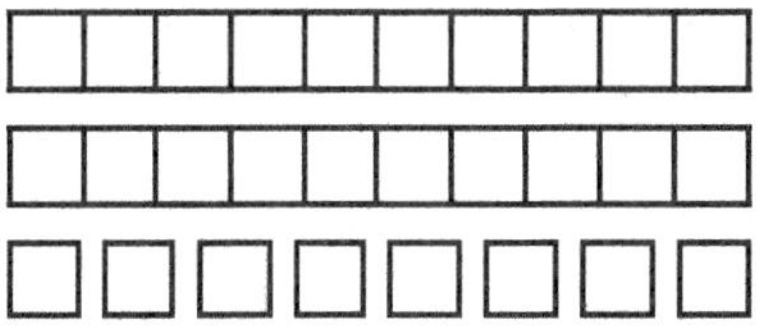

______

____ is __________ than 25.

18.

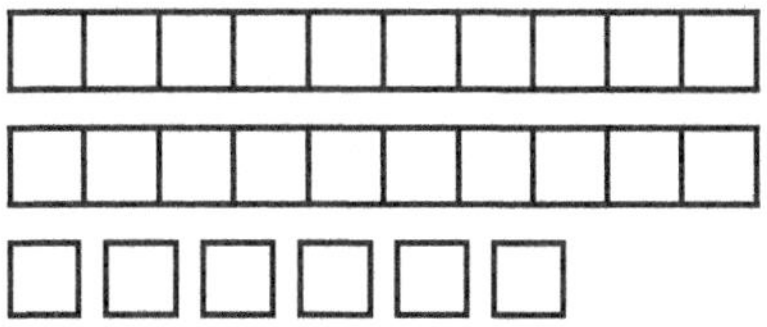

______

____ is __________ than 25.

19.

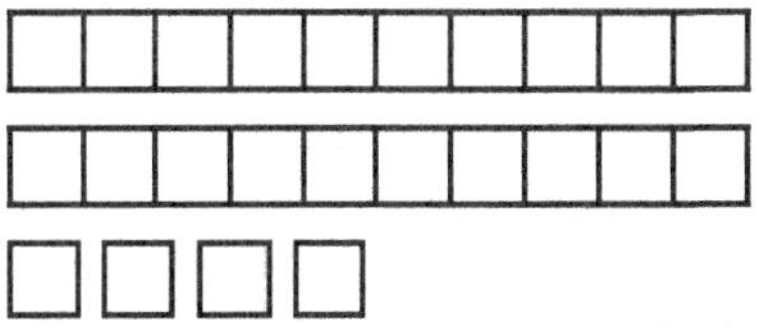

______

____ is __________ than 25.

Comparing Numbers Up to 100

☐ Circle the largest number.

1.

3

(9)

5

2.

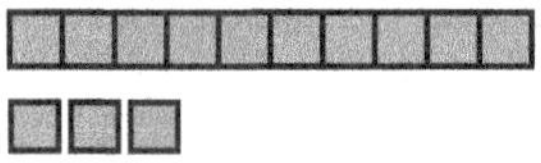
13

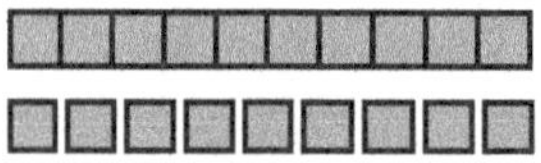
19

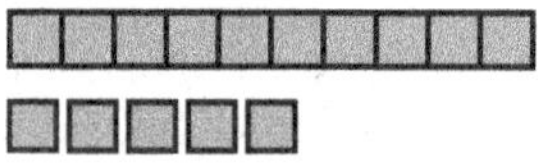
15

3.

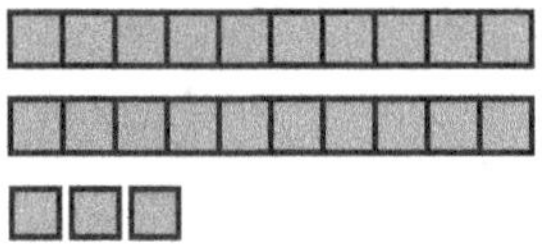
23

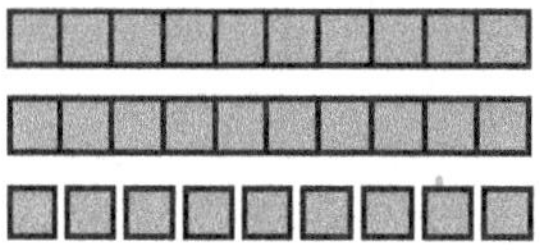
29

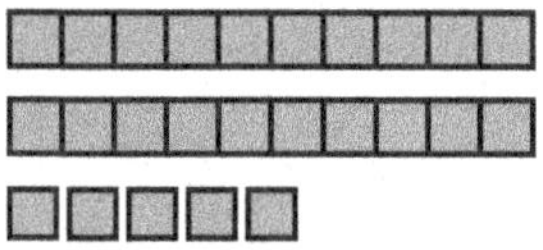
25

4. 43 49 45

5. 73 75 79

☐ Write the numbers in order from smallest to largest.

6. 4 7 6

4 ______ ______ ______

7. 34 37 36

______ ______ ______

8. 49 43 44

______ ______ ______

9. 82 80 85

______ ______ ______

10. BONUS

Circle the two numbers that are not in order.

51 54 55 58 57 59

☐ Circle the largest number.

11.

14

4

24

12.

16

6

26

13.

13

3

23

14.

15 25 5

15.

34 74 54

16.

29 19 89

☐ Write the numbers in order from smallest to largest.

17.

53 63 43

43 ______ ______

18.

27 7 37

______ ______ ______

19. BONUS

Circle the two numbers that are not in order.

24 34 44 84 74 94

☐ Circle the larger number.

20.

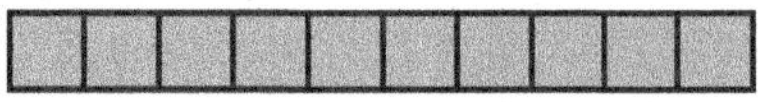

10

9

21.

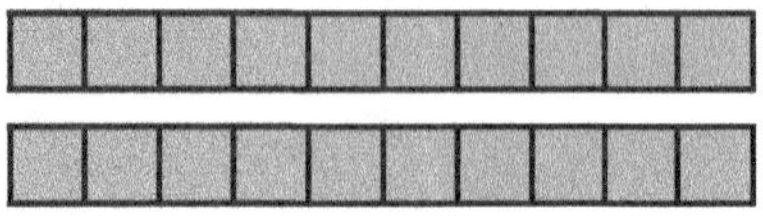

20

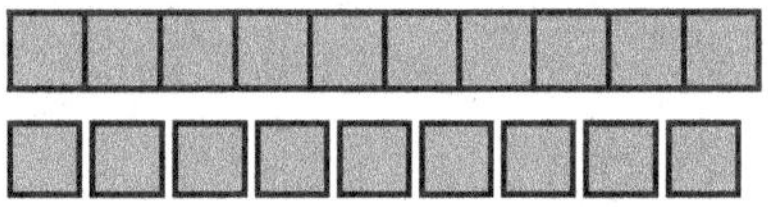

19

22. 30 29

23. 49 50

24. 80 79

25. 39 40

26. 60 59

27. 90 89

28.

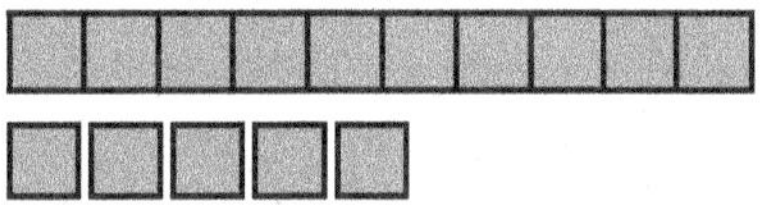

15

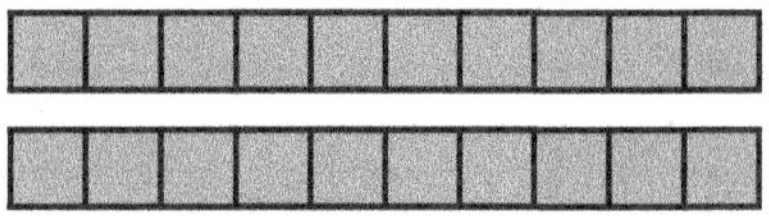

20

29.

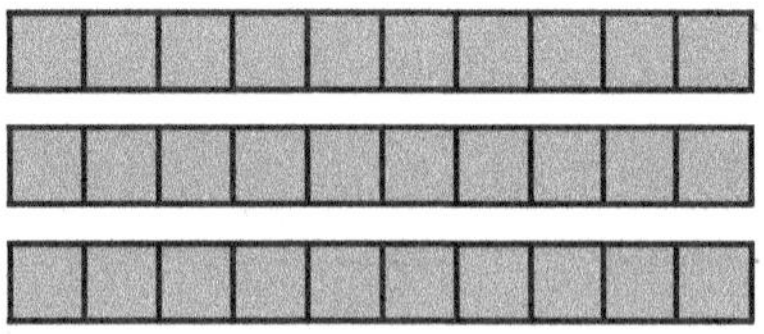

30

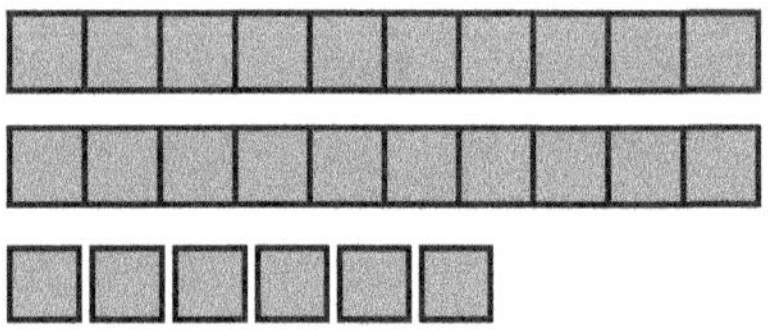

26

30. 47 50

31. 60 53

32. 80 90

NBT1-19 The Hundreds Chart

☐ Color all the numbers with ones digit 4 blue.

☐ Color all the numbers with tens digit 3 yellow.

1.

1	2	3	4	5	6	7	8	9	10
11	12	13	14	15	16	17	18	19	20
21	22	23	24	25	26	27	28	29	30
31	32	33	34	35	36	37	38	39	40
41	42	43	44	45	46	47	48	49	50
51	52	53	54	55	56	57	58	59	60
61	62	63	64	65	66	67	68	69	70
71	72	73	74	75	76	77	78	79	80
81	82	83	84	85	86	87	88	89	90
91	92	93	94	95	96	97	98	99	100

☐ Use the chart to answer the questions.

2. What color is 30? ________________

3. What color is 34? ________________

☐ Color all the numbers with tens digit 7 red.

☐ Color all the numbers with ones digit 2 blue.

4.

1	2	3	4	5	6	7	8	9	10
11	12	13	14	15	16	17	18	19	20
21	22	23	24	25	26	27	28	29	30
31	32	33	34	35	36	37	38	39	40
41	42	43	44	45	46	47	48	49	50
51	52	53	54	55	56	57	58	59	60
61	62	63	64	65	66	67	68	69	70
71	72	73	74	75	76	77	78	79	80
81	82	83	84	85	86	87	88	89	90
91	92	93	94	95	96	97	98	99	100

☐ Use the chart to answer the questions.

5.

What color is 22? ______________

6.

What number is red and blue? ______________

↑ up ↓ down right → left ←

☐ Draw an arrow to show the direction you move on a hundreds chart.

7.

←	47	

to find 41

	↑	
	47	

to find 27

	47	→

to find 49

	47	
	↓	

to find 97

8.

	53	

to find 93

	53	

to find 58

	53	

to find 51

	53	

to find 23

9.

	25	

to find 21

	25	

to find 29

	25	

to find 45

	25	

to find 65

10.

	64	

to find 69

	64	

to find 84

	64	

to find 24

	64	

to find 62

NBT1-20 Ordering Numbers Up to 100

☐ Write the numbers in order from smallest to largest.

1.

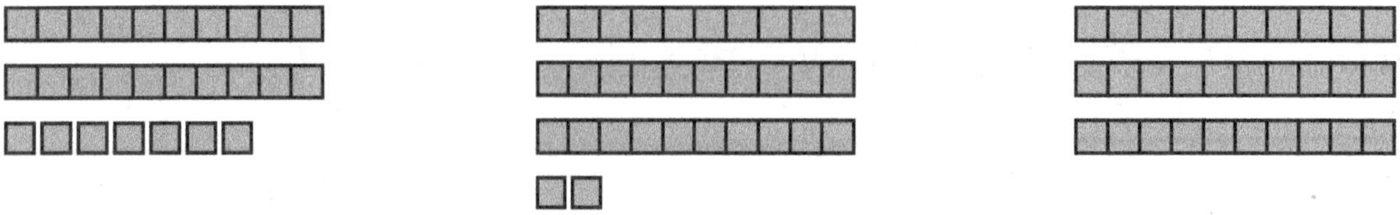

27 32 30

_____ _____ _____

2.

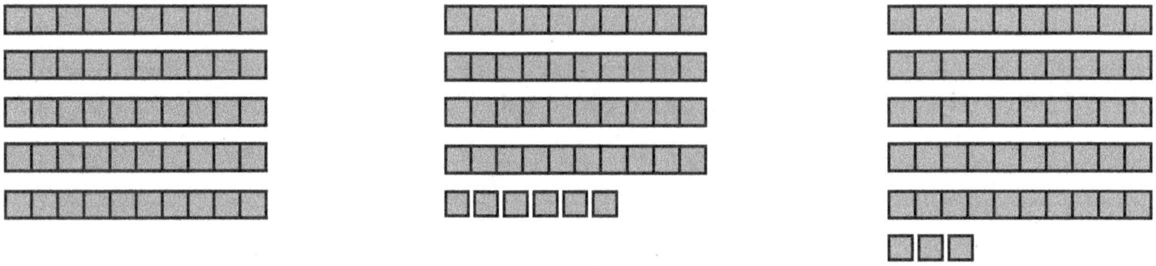

50 46 53

_____ _____ _____

3. 57 61 50

_____ _____ _____

4. 23 18 20

_____ _____ _____

Zara tries to write the numbers from smallest to largest.

☐ Circle Zara's mistake.

5. 23 25 34 30 41

6. 54 73 68 75 80

☐ Write the shaded numbers in order.

7.

31	32	**33**	34	35	36	**37**	38	39	**40**
41	42	**43**	44	45	46	47	48	49	**50**
51	52	53	54	55	56	57	**58**	59	60

33 ______ ______ ______ ______ ______ ______

8.

51	**52**	53	54	**55**	56	57	**58**	59	60
61	62	63	64	65	**66**	67	68	69	70
71	72	**73**	**74**	75	76	77	78	79	80

______ ______ ______ ______ ______ ______ ______

9.

71	72	73	**74**	75	**76**	77	78	79	80
81	82	**83**	84	85	86	87	88	**89**	90
91	92	93	94	95	96	**97**	98	**99**	100

______ ______ ______ ______ ______ ______ ______

☐ Look at a meter stick to check your answers.

☐ Shade the numbers.
☐ Circle the numbers as you shade them.
☐ Write the numbers in order.

10.

(41) 38 22 35 29 40 47

21	22	23	24	25	26	27	28	29	30
31	32	33	34	35	36	37	38	39	40
41	42	43	44	45	46	47	48	49	50

____ ____ ____ ____ ____ ____ ____

11.

59 42 63 56 45 50 61

41	42	43	44	45	46	47	48	49	50
51	52	53	54	55	56	57	58	59	60
61	62	63	64	65	66	67	68	69	70

____ ____ ____ ____ ____ ____ ____

NBT1-21 Comparing Numbers Using >

The > sign means **greater than**.

☐ Fill in the number sentence.

1. 7 is greater than 3.
   7 > 3

2. 19 is greater than 8.
   ____ > ____

3. 72 is greater than 41.
   ____ > ____

4. 37 is greater than 4.
   ____ > ____

5. 62 is greater than 26.
   ____ > ____

6. 55 is greater than 47.
   ____ > ____

☐ Circle **yes** or **no**.

7. 87 > 19 — (yes) no
8. 17 > 21 — yes no
9. 36 > 31 — yes no
10. 57 > 50 — yes no
11. 42 > 91 — yes no
12. 51 > 16 — yes no
13. 61 > 78 — yes no
14. 40 > 60 — yes no
15. 92 > 8 — yes no
16. 23 > 41 — yes no
17. 35 > 16 — yes no
18. 53 > 68 — yes no

☐ Circle the greater number.

☐ Write the numbers in the correct places.

19. (21) 17

21 > 17

20. 32 (57)

57 > 32

21. 8 90

_____ > _____

22. 15 80

_____ > _____

23. 33 2

_____ > _____

24. 42 81

_____ > _____

25. 61 30

_____ > _____

26. 85 96

_____ > _____

27. 77 52

_____ > _____

☐ Circle the greater amount.

☐ Write the numbers in the correct places.

28.

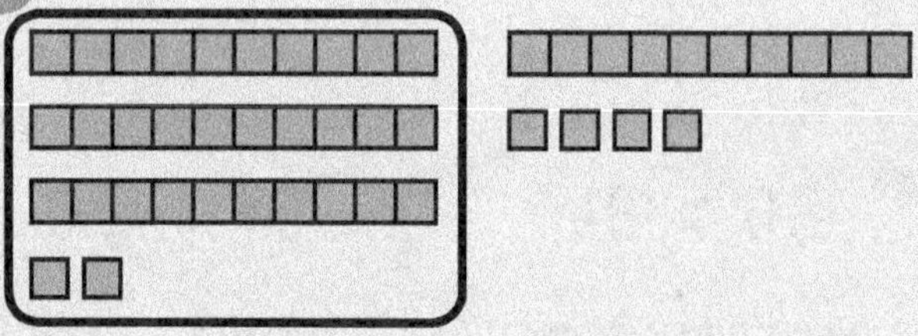

32 > 14

29.

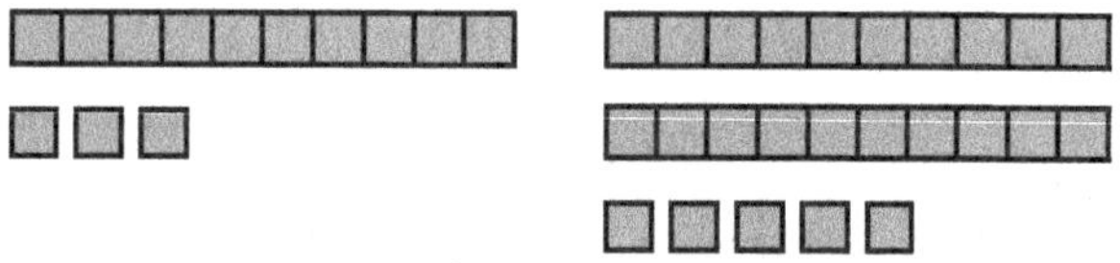

_____ > _____

30.

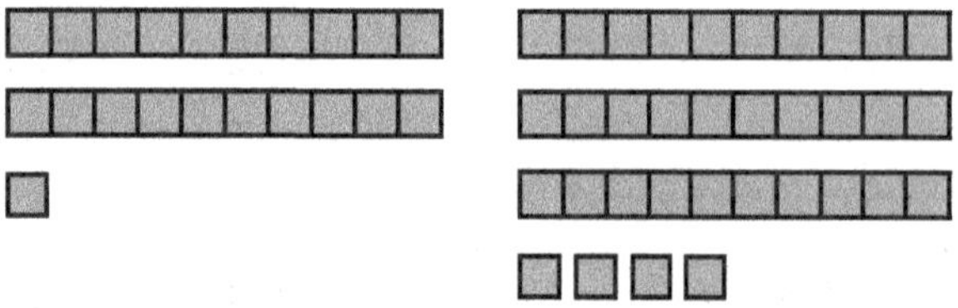

_____ > _____

31.

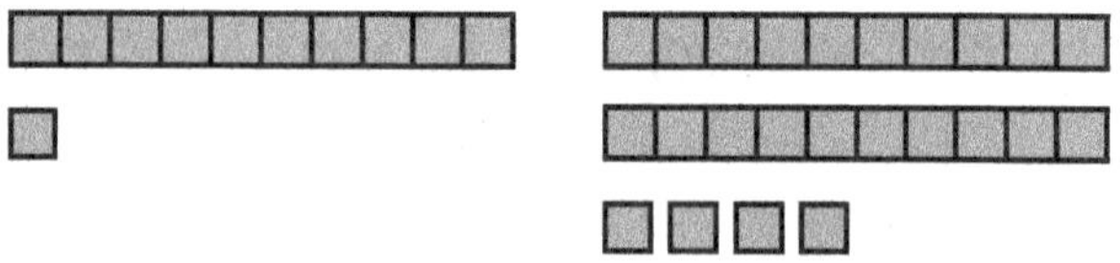

_____ > _____

NBTI-22 Comparing Numbers Using <

The < sign means **less than**.

☐ Fill in the number sentence.

1. 12 is less than 32.
   12 < 32

2. 19 is less than 97.
   ___ < ___

3. 18 is less than 58.
   ___ < ___

4. 82 is less than 90.
   ___ < ___

5. 67 is less than 74.
   ___ < ___

6. 59 is less than 65.
   ___ < ___

☐ Circle **yes** or **no**.

7. $82 < 21$ yes (no)
8. $91 < 97$ yes no
9. $18 < 68$ yes no
10. $23 < 16$ yes no
11. $40 < 49$ yes no
12. $91 < 95$ yes no
13. $82 < 17$ yes no
14. $78 < 12$ yes no
15. $6 < 11$ yes no
16. $2 < 60$ yes no
17. $35 < 9$ yes no
18. $82 < 43$ yes no

☐ Circle the number that is less.

☐ Write the numbers in the correct places.

19. (20) 30
20 < 30

20. 19 3
_____ < _____

21. 27 65
_____ < _____

22. 92 83
_____ < _____

23. 41 56
_____ < _____

24. 52 13
_____ < _____

25. 20 50
_____ < _____

26. 73 21
_____ < _____

27. 2 7
_____ < _____

☐ Circle the smaller amount.

☐ Write the numbers in the correct places.

28.
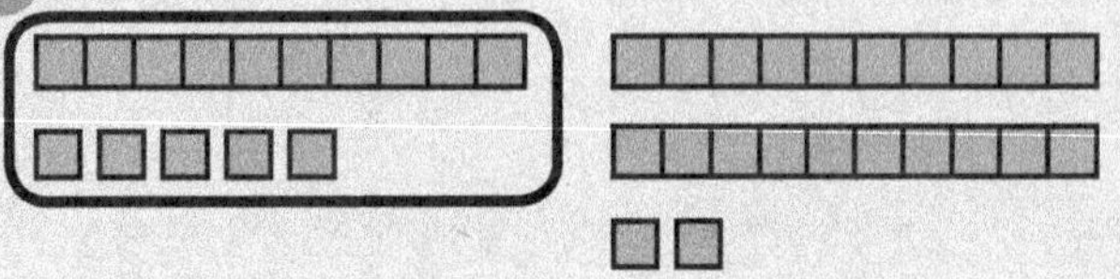
15 < 22

29.
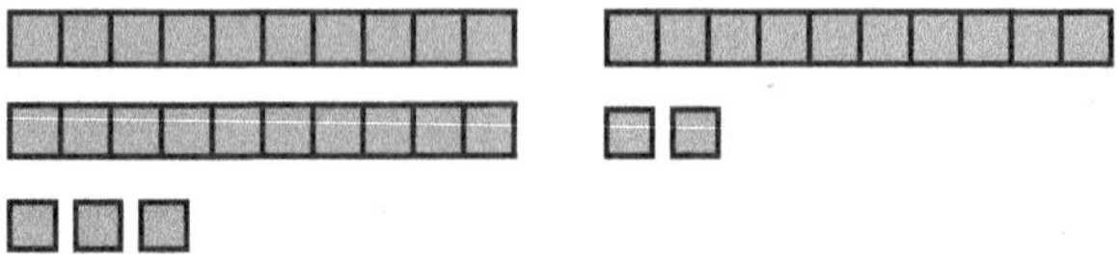
_____ < _____

30.
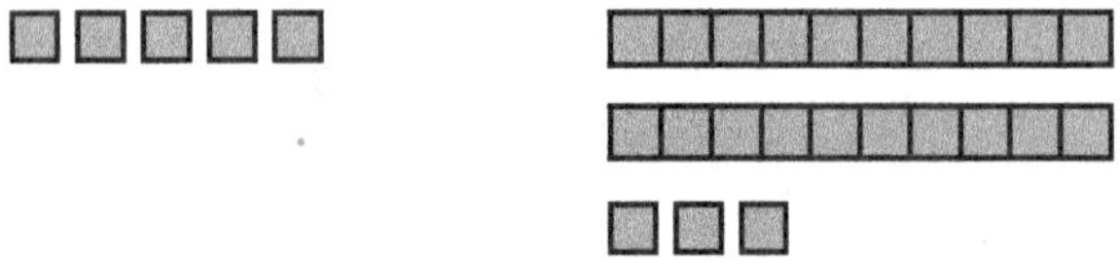
_____ < _____

31.
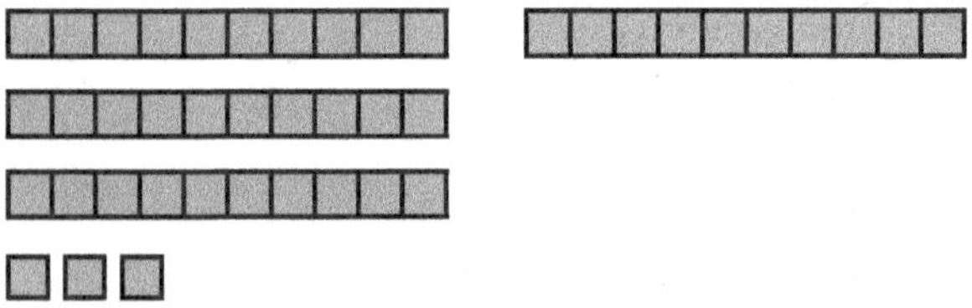
_____ < _____

NBTI-23 Comparing Numbers Using > and <

> means **greater than.** < means **less than.**

☐ Circle the correct sign.

1. 21 (>, <) 27 — < circled	2. 32 (>, <) 12 — > circled	3. 84 (>, <) 13	4. 63 (>, <) 71
5. 50 (>, <) 70	6. 80 (>, <) 20	7. 36 (>, <) 94	8. 14 (>, <) 47
9. 46 (>, <) 3	10. 15 (>, <) 35	11. 72 (>, <) 89	12. 63 (>, <) 64

☐ Circle **yes** or **no**.

13. 17 > 23 yes no — no circled	14. 65 > 18 yes no	15. 32 < 77 yes no	16. 41 > 68 yes no
17. 73 > 89 yes no	18. 93 > 14 yes no	19. 11 > 3 yes no	20. 8 > 29 yes no
21. 25 < 83 yes no	22. 14 < 69 yes no	23. 97 > 49 yes no	24. 44 < 13 yes no

☐ Circle the greater number.

☐ Write the numbers in the correct places.

25. 9 (18)

18 > 9

9 < 18

26. 92 16

____ > ____

____ < ____

27. 7 35

____ > ____

____ < ____

28. 64 9

____ > ____

____ < ____

29. 30 45

____ > ____

____ < ____

30. 23 91

____ > ____

____ < ____

31. 83 75

____ > ____

____ < ____

32. 17 65

____ > ____

____ < ____

33. 32 37

____ > ____

____ < ____

34. 40 50

____ > ____

____ < ____

35. 60 38

____ > ____

____ < ____

36. 53 94

____ > ____

____ < ____

37. 13 30

____ > ____

____ < ____

38. 84 38

____ > ____

____ < ____

39. 14 44

____ > ____

____ < ____

☐ Write the numbers in the correct places.

40. 24 26

___ > ___

41. 52 63

___ > ___

42. 90 36

___ < ___

43. 32 45

___ < ___

44. 74 83

___ < ___

45. 19 7

___ > ___

46. 16 7

___ < ___

47. 85 60

___ > ___

48. 97 2

___ < ___

☐ Circle the greater number.

☐ Write the numbers in the correct places.

49.

Tens	Ones
5	2
1	8

52 > 18

50.

Tens	Ones
9	1
4	3

___ > ___

51.

Tens	Ones
1	5
3	5

___ > ___

52.

Tens	Ones
2	1
3	9

___ < ___

53.

Tens	Ones
7	6
7	2

___ < ___

54.

Tens	Ones
6	2
5	0

___ > ___

NBTI-24 Greater Than, Less Than, and Equal

☐ Write the numbers in the blanks.

☐ Write >, <, or = in the circle.

1.

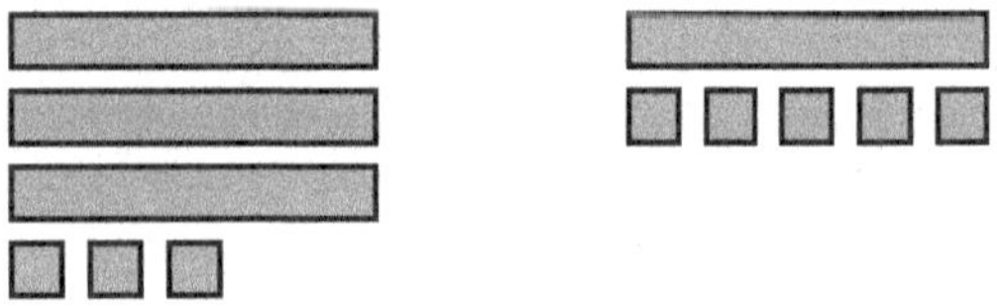

33 ◯ 15

2.

_____ ◯ _____

3.

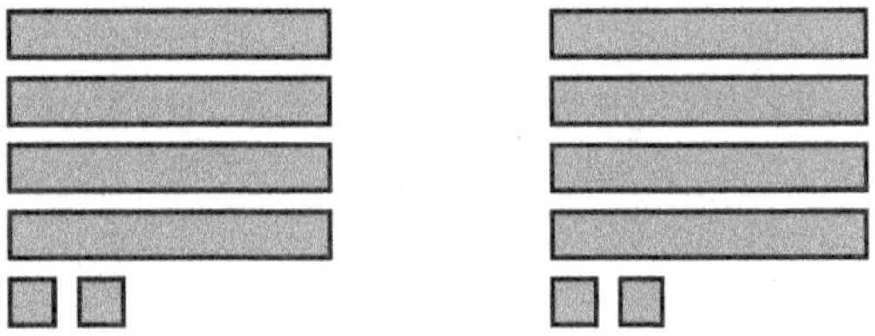

_____ ◯ _____

4.

_____ ◯ _____

5.

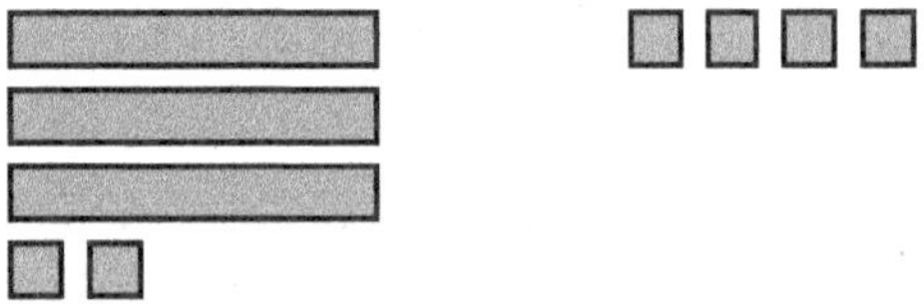

_____ ◯ _____

6.

_____ ◯ _____

7.

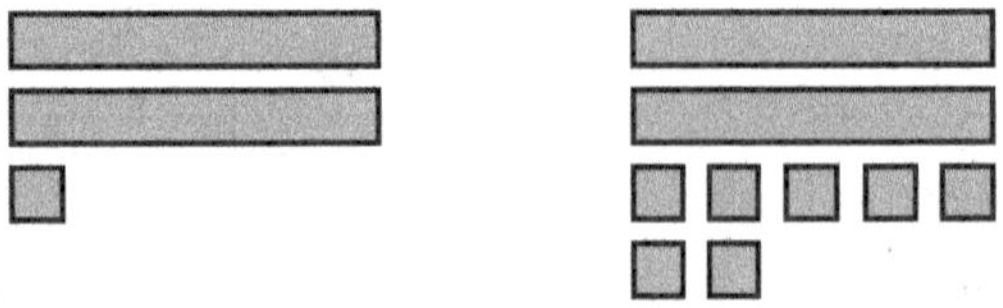

_____ ◯ _____

8.

_____ ◯ _____

☐ Write the numbers in the blanks.

☐ Write >, <, or = in the circle.

9.

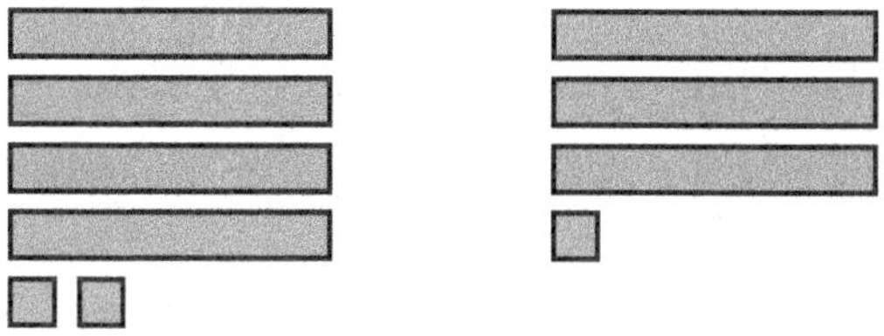

42 ◯ 31

10.

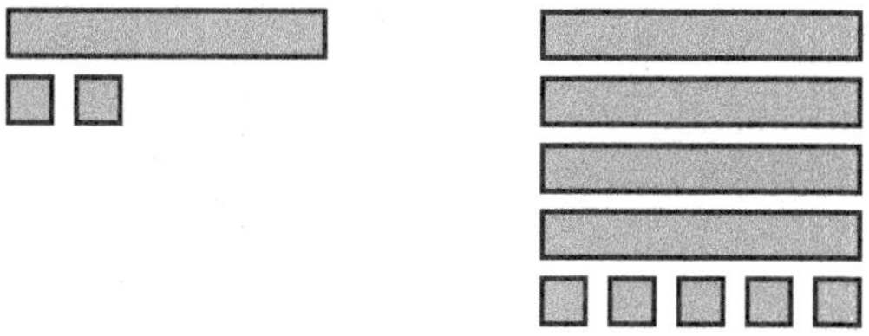

______ ◯ ______

11.

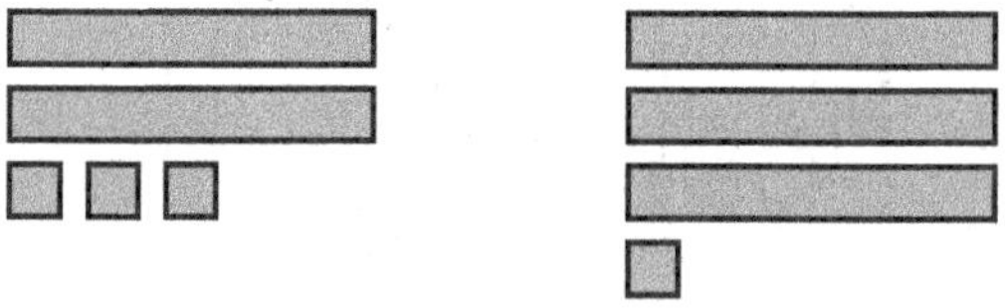

______ ◯ ______

12.

______ ◯ ______

13.

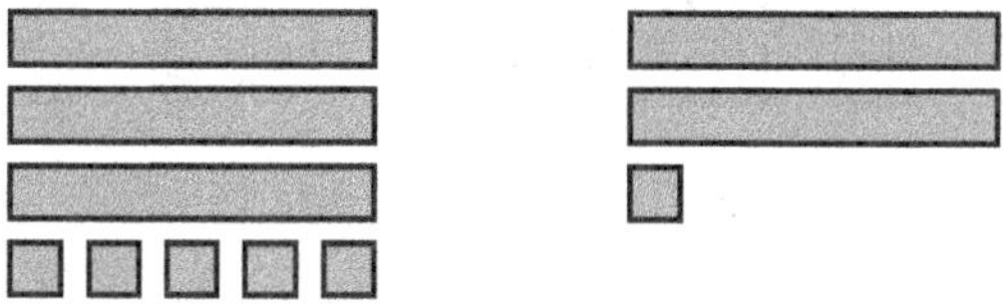

______ ◯ ______

14. BONUS

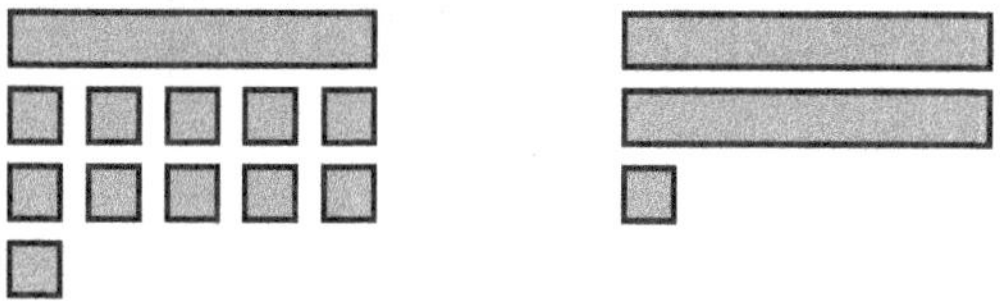

______ ◯ ______

15. Explain how you know 82 is greater than 72.

16. Explain how you know 54 is less than 59.

NBTI-25 Counting to 109

100 is 10 tens.

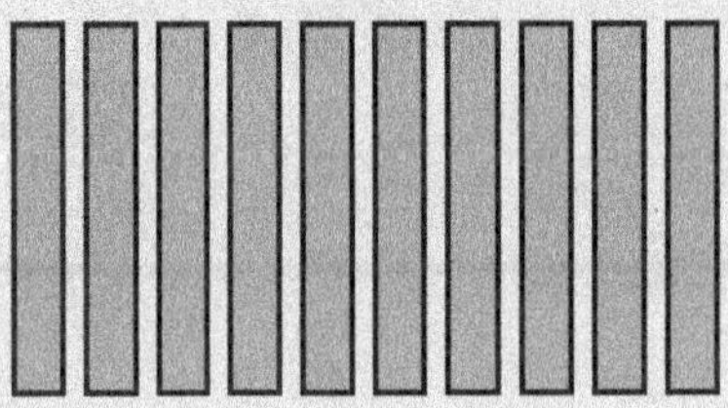

☐ Write the number.

1.

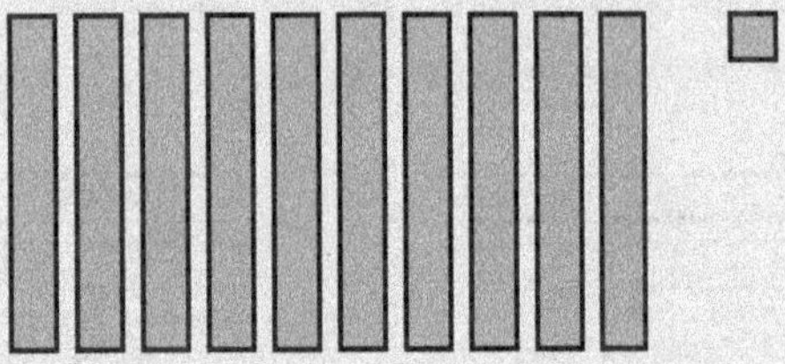

101

2.

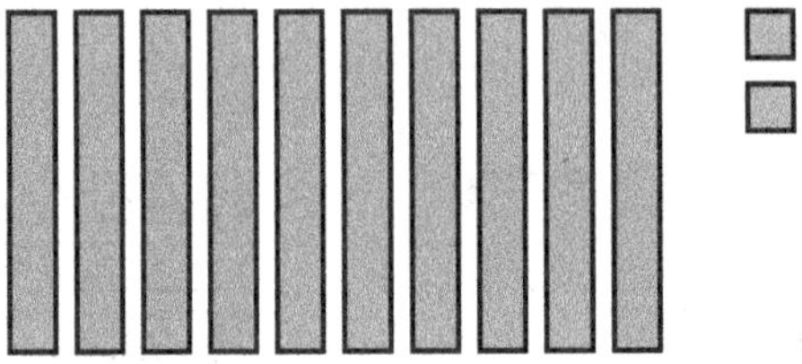

_____

3.

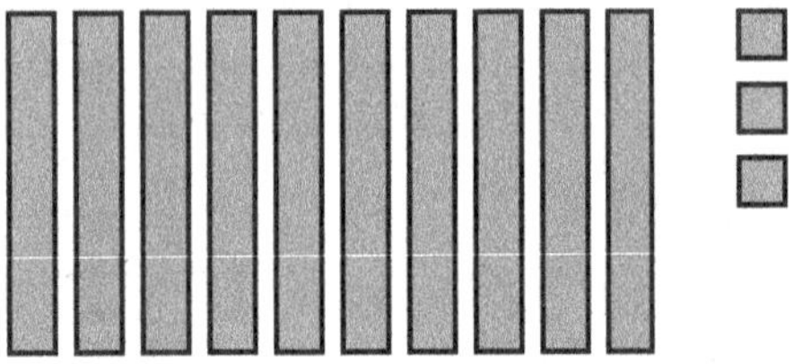

_____

4.

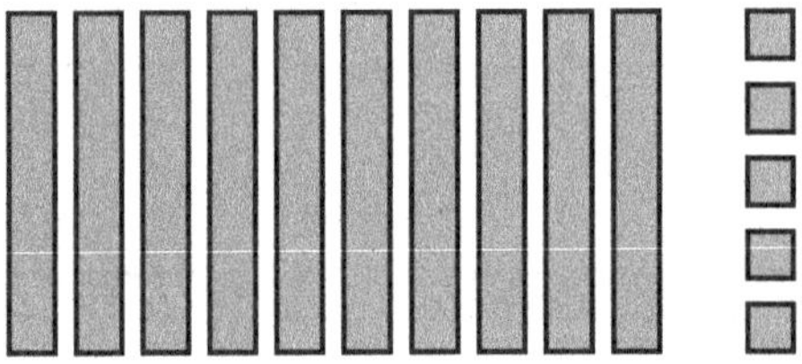

_____

5.

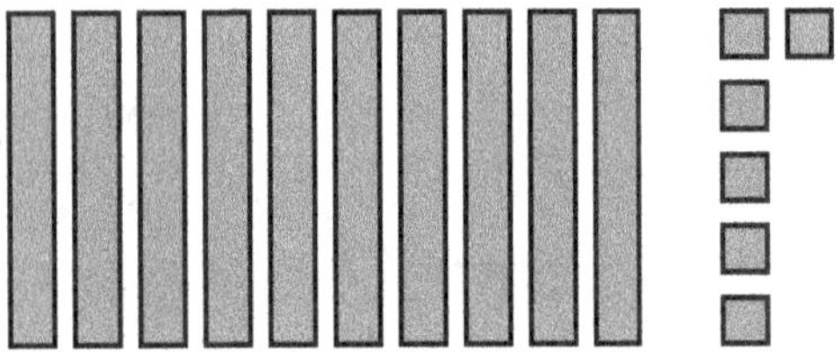

_____

6.

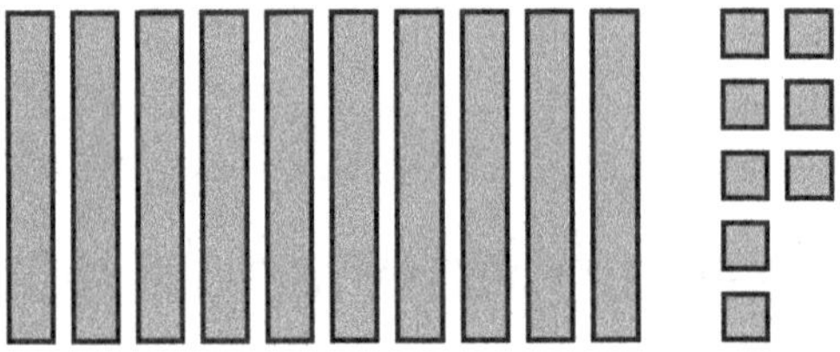

_____

☐ Add.

7. $100 + 1 =$ ____

8. $100 + 2 =$ ____

9. $100 + 7 =$ ____

10. $100 + 9 =$ ____

11. $100 + 8 =$ ____

12. $100 + 3 =$ ____

13. $100 + 6 =$ ____

14. $100 + 5 =$ ____

15. $100 + 4 =$ ____

☐ Fill in the blanks.

16. 100, 101, 102, ____, ____, ____

17. 97, 98, ____, ____, ____, ____

18. 104, ____, 106, ____, ____, ____

19. Make 107 using base ten blocks.

20. Andy has 102 stamps.
Make this number using base ten blocks.

NBTI-26 Counting to 120

120 is 11 tens.

☐ Write the number.

1\.

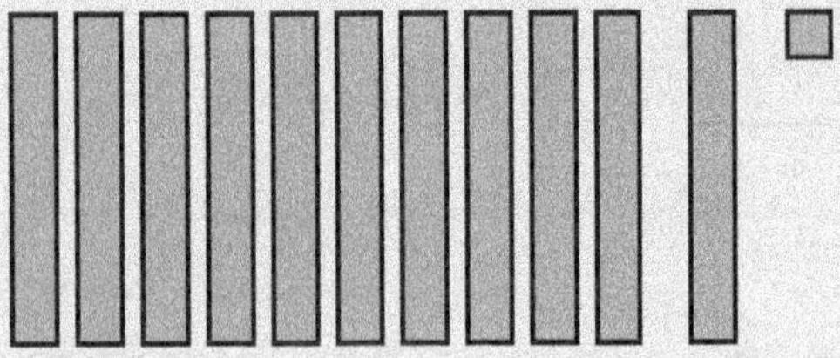

111

2\.

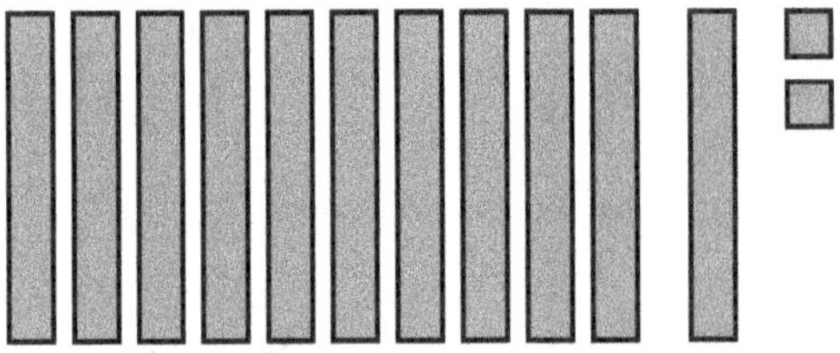

______

3\.

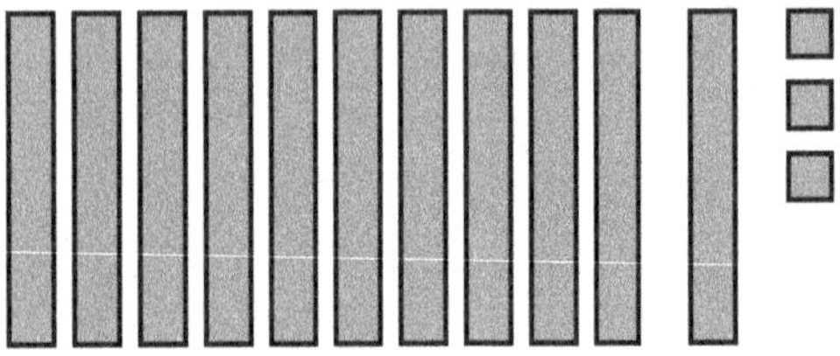

______

4\.

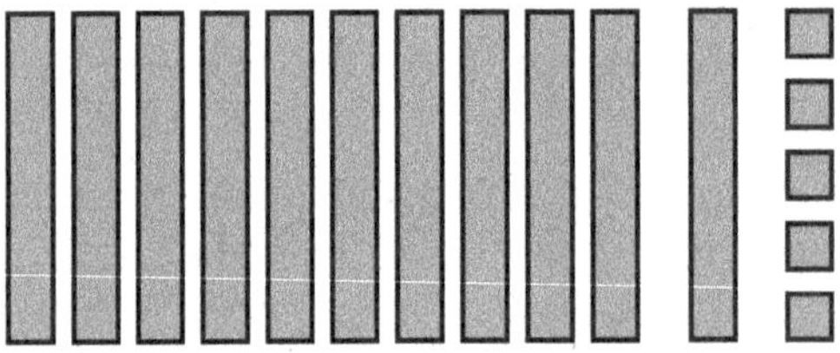

______

5\.

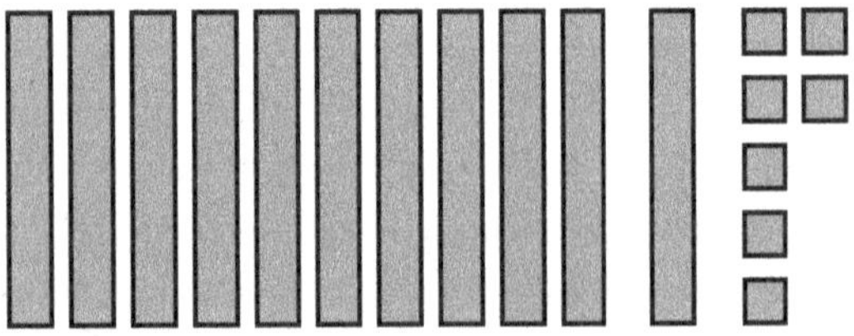

______

6\. BONUS

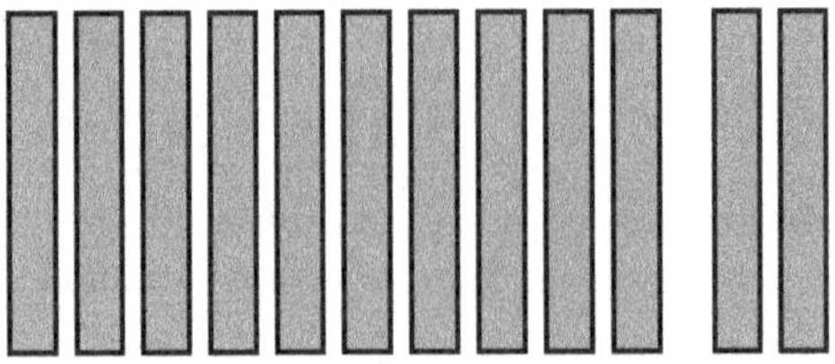

______

☐ Add.

7. 110 + 1 = _____

8. 110 + 3 = _____

9. 110 + 4 = _____

10. 110 + 7 = _____

11. 110 + 1 = _____

12. 110 + 9 = _____

13. 110 + 2 = _____

14. 110 + 5 = _____

15. 110 + 8 = _____

16. 100 + 2 = _____
110 + 2 = _____

17. 100 + 3 = _____
110 + 3 = _____

18. 100 + 9 = _____
110 + 9 = _____

19. 100 + 7 = _____
110 + 7 = _____

20. 100 + 6 = _____
110 + 6 = _____

21. BONUS
100 + 14 = _____

☐ Fill in the blanks.

22. 110, 111, _____, _____, _____, _____

23. 115, 116, _____, _____, _____, _____

24. 105, _____, _____, _____, 109, _____

☐ Circle the picture that shows the number.

25. 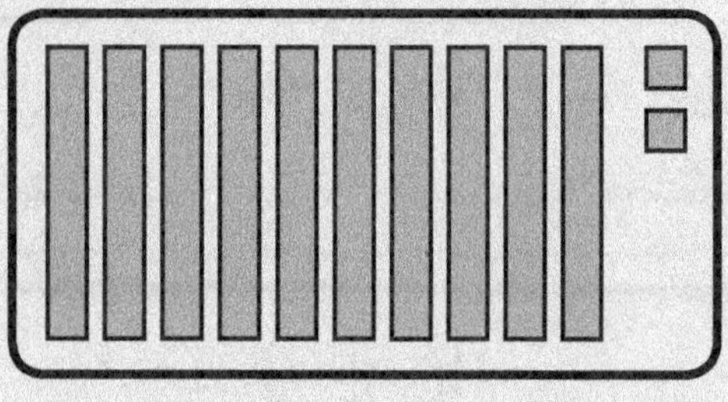 102 

26. 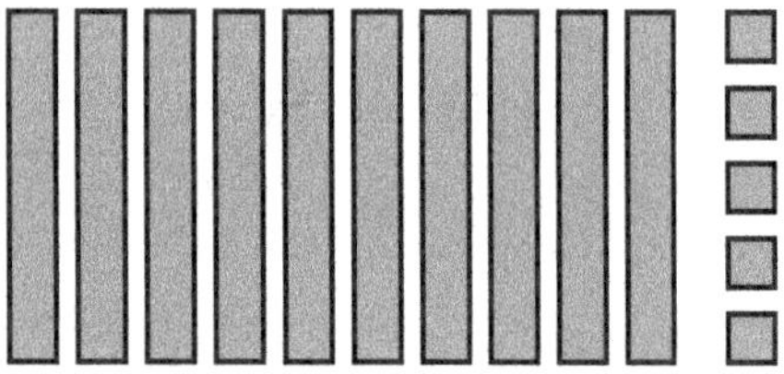 115 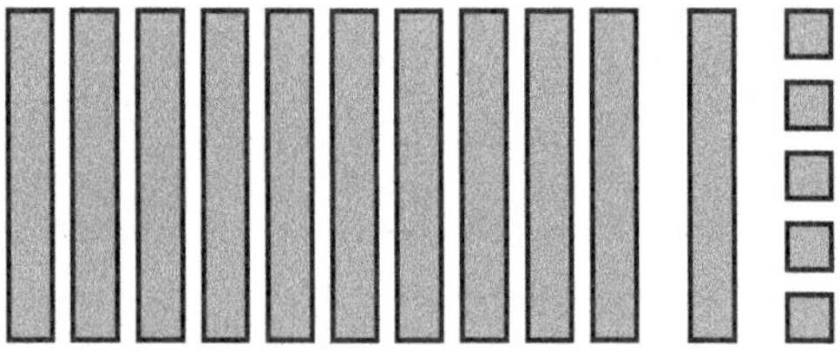

27. 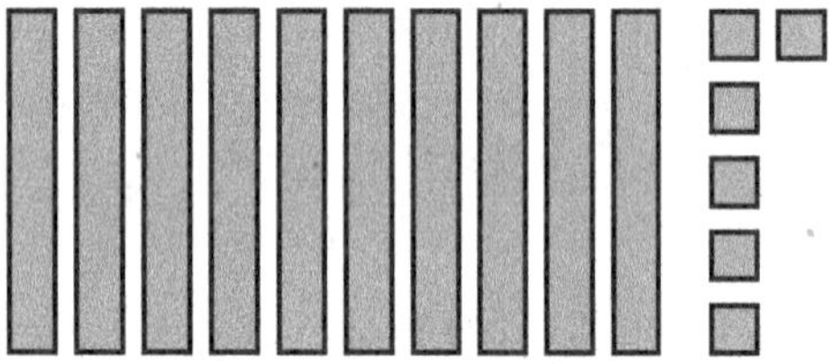 106 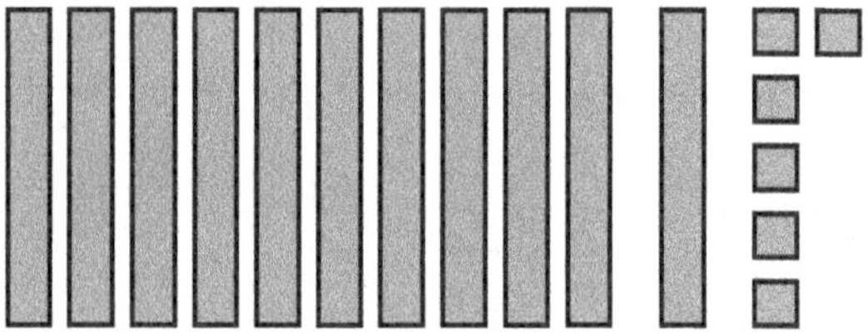

28. 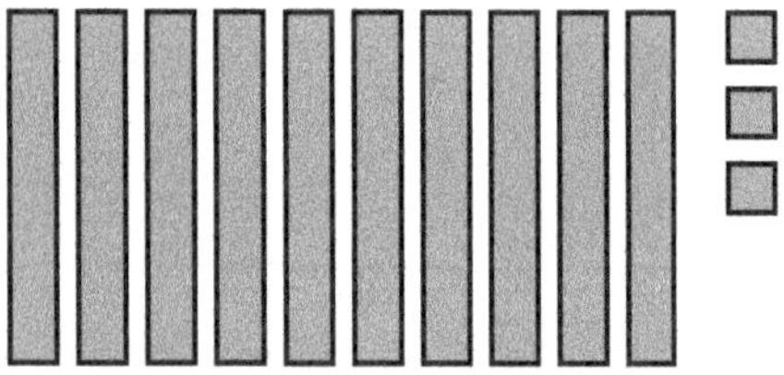 113 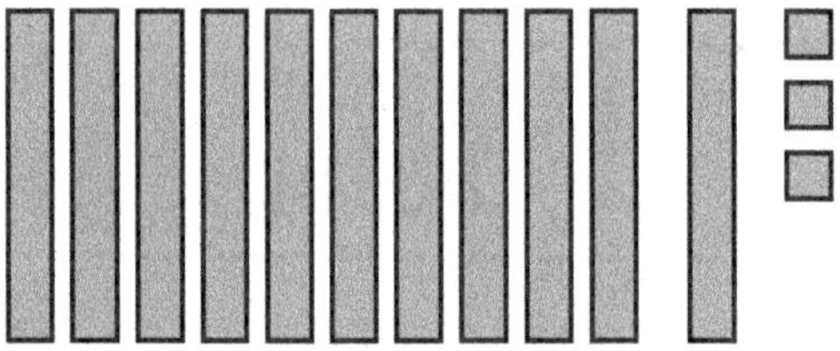

29. Anne says the picture shows 105.

Explain her mistake.

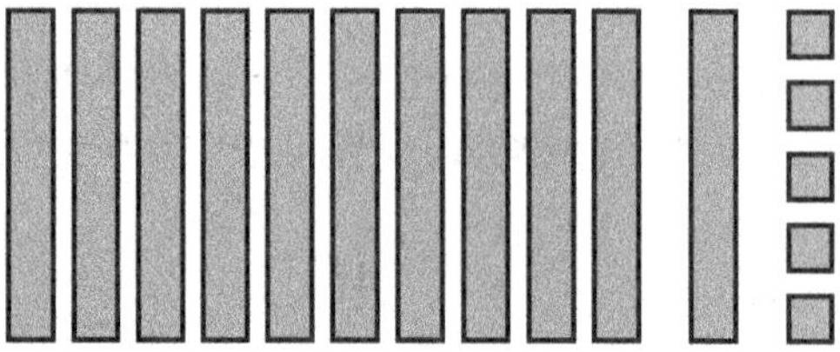

NBT1-27 Review—Comparing and Ordering

☐ Fill in the missing numbers.

1.

41		43			46			49	50
51				55		57			
			64		66		68		70

2.

81	82			85		87			90
	92		94			97			100
101	102		104	105			108		

3.

				95		97			100
	102			105		107	108		
111			114		116				120

☐ Write >, <, or =.

4. 27 ◯ 32

5. 92 ◯ 92

6. 18 ◯ 36

7. 99 ◯ 76

8. 11 ◯ 57

9. 81 ◯ 62

10. 85 ◯ 62

11. 43 ◯ 34

12. 12 ◯ 12

☐ Write the numbers in order.

13. 27 19 3 82 56

____ ____ ____ ____ ____

14. 43 11 50 9 34

____ ____ ____ ____ ____

15. Ken has 37 stickers.
Liz has 25 stickers.
Who has more stickers?

16. Sun's puppy is 90 days old.
Mike's puppy is 87 days old.
Whose puppy is older?

NBTI-28 Adding Tens to Tens

☐ Write the additions for the picture.

1.

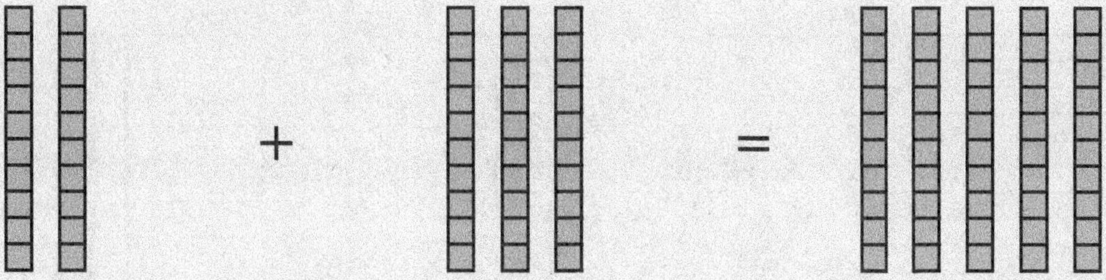

__2__ tens + __3__ tens = __5__ tens

__20__ + __30__ = __50__

2.

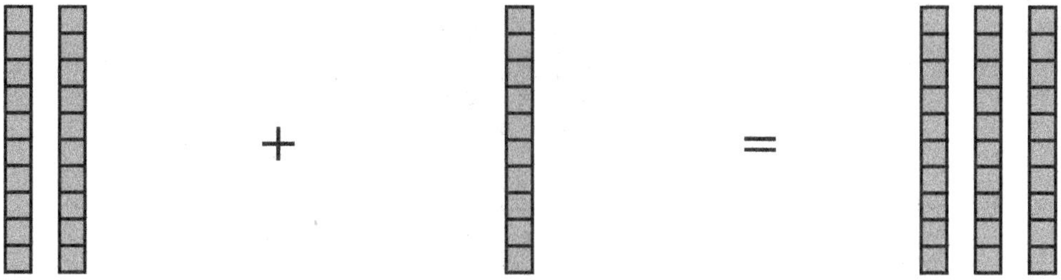

_____ tens + _____ ten = _____ tens

_____ + _____ = _____

3.

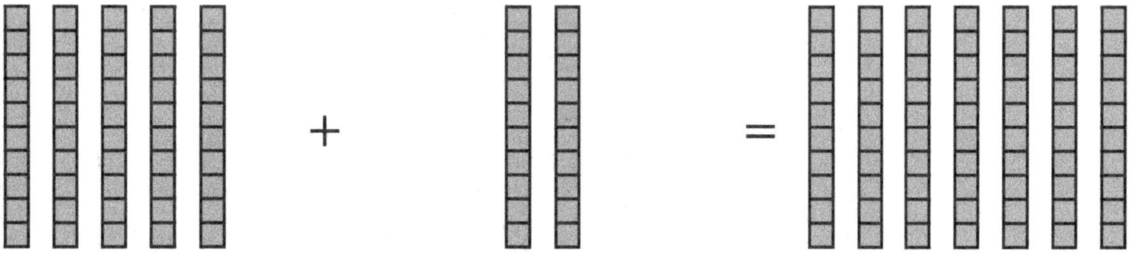

_____ tens + _____ tens = _____ tens

_____ + _____ = _____

☐ Draw tens to add.

☐ Write the additions for the picture.

4.

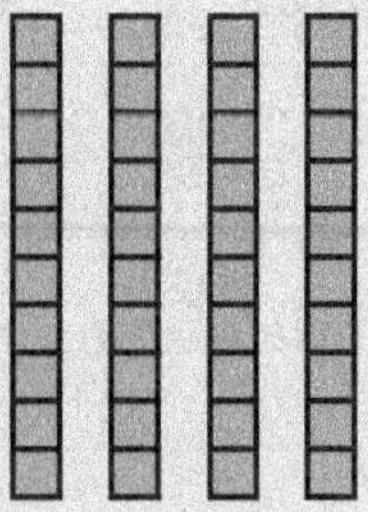 + 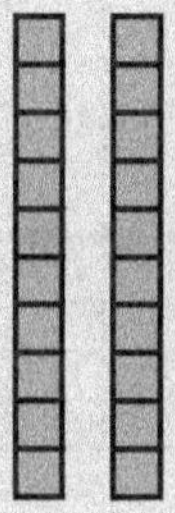 = 

__4__ tens + __2__ tens = __6__ tens

__40__ + __20__ = __60__

5.

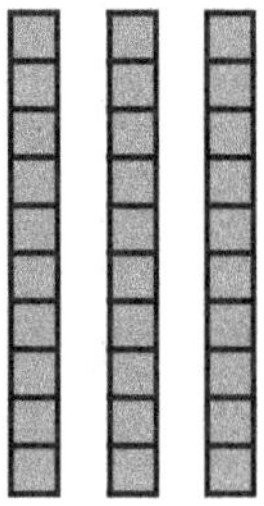 + 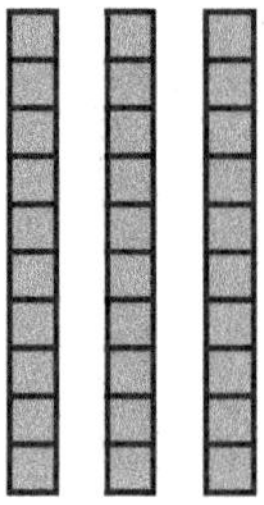 =

_____ tens + _____ tens = _____ tens

_____ + _____ = _____

6.

= 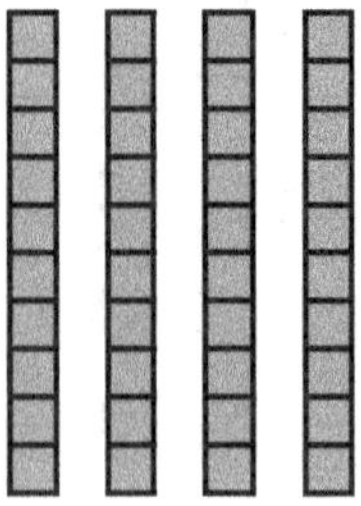 + 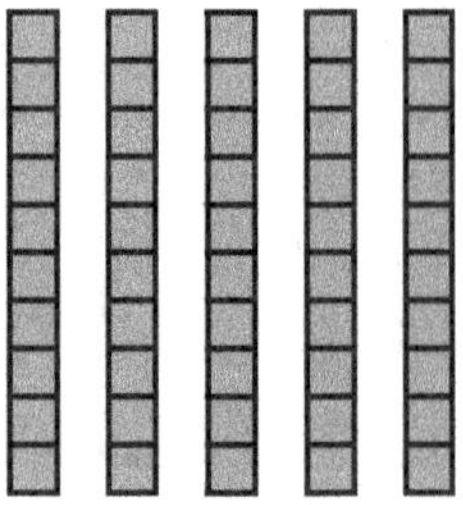

_____ tens = _____ tens + _____ tens

_____ = _____ + _____

Add.

7.

5 ones	5	5 tens	50
+ 2 ones	+ 2	+ 2 tens	+ 20
7 ones	7	7 tens	70

8.

4 ones	4	4 tens	40
+ 3 ones	+ 3	+ 3 tens	+ 30
☐ ones	☐	☐ tens	☐

9.

6 ones	6	6 tens	60
+ 2 ones	+ 2	+ 2 tens	+ 20
☐ ones	☐	☐ tens	☐

10.

3	30
+ 6	+ 60
☐	☐

11.

1	10
+ 4	+ 40
☐	☐

12.

5	50
+ 4	+ 40
☐	☐

13.

8 + 1 = _____

80 + 10 = _____

14.

2 + 7 = _____

20 + 70 = _____

15.

6 + 3 = _____

60 + 30 = _____

☐ Draw tens to find the missing number.

16.

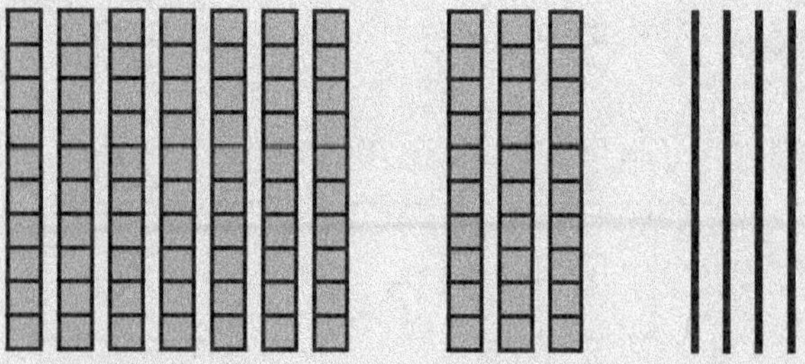

70 = 30 + 40

17.

60 = 40 + ____

18.

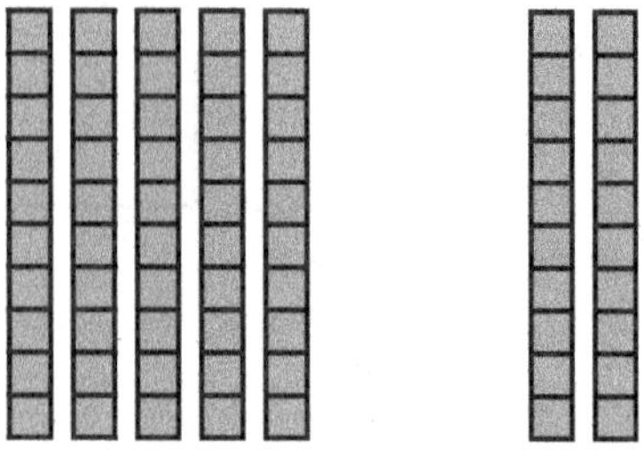

50 = 20 + ____

19.

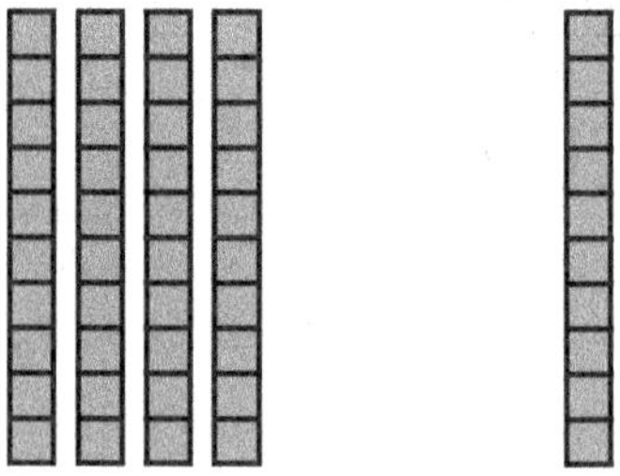

40 = 10 + ____

☐ Find the missing numbers.

20.

7 = 5 + ____

70 = 50 + ____

21.

8 = 6 + ____

80 = 60 + ____

22.

5 = 4 + ____

50 = 40 + ____

☐ Draw tens to solve the problem.

23.

Anwar has 20 bus tickets.
May has 30 bus tickets.
How many tickets do they have altogether?

NBT1-29 Subtracting Tens from Tens

☐ Cross out tens to subtract.

1.

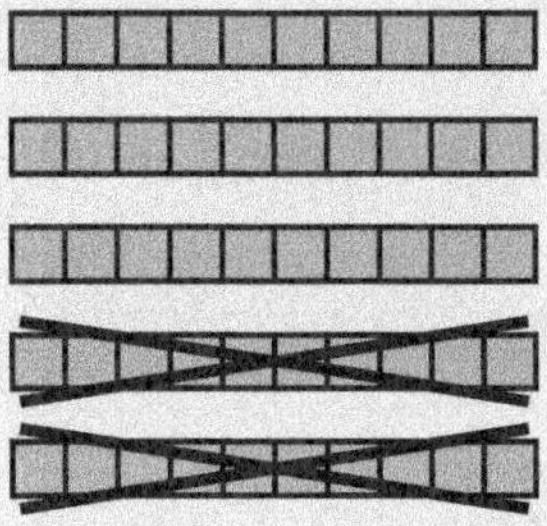

50 − 20 = 30

2.

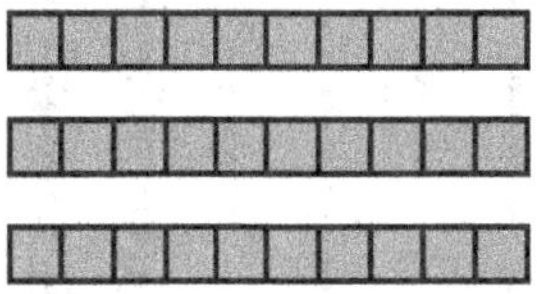

30 − 10 = ______

3.

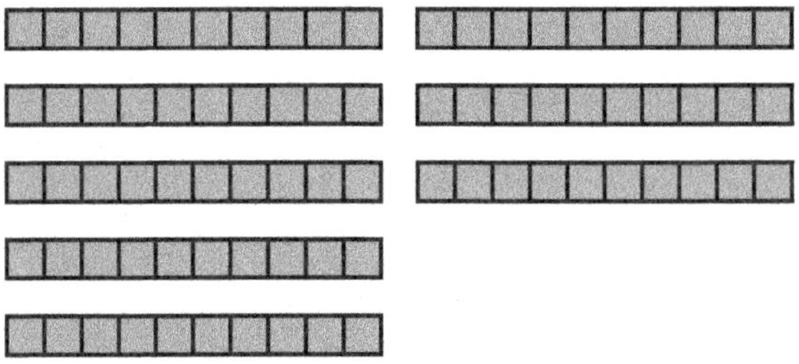

80 − 30 = ______

4.

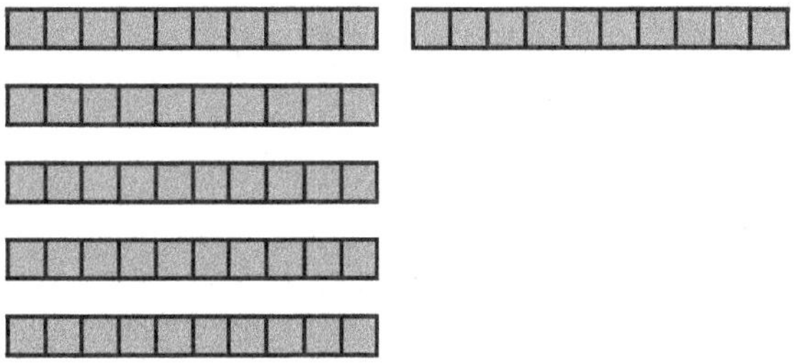

60 − 40 = ______

5.

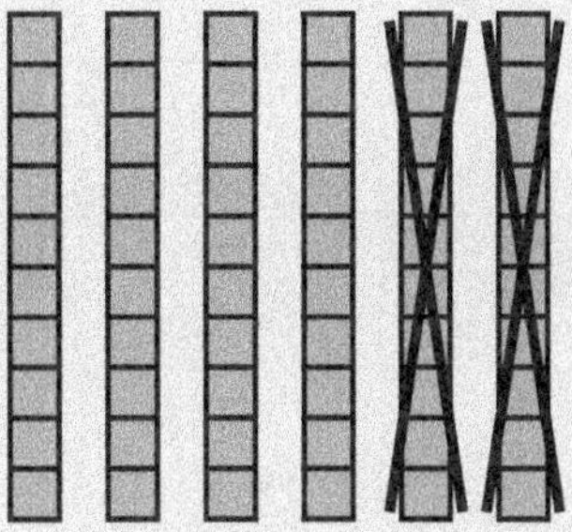

60 − 20 = 40

6.

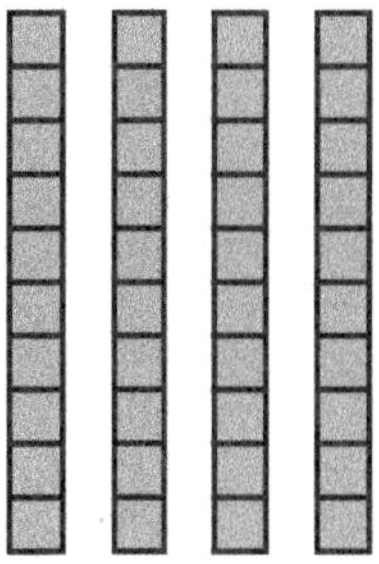

40 − 10 = ______

7.

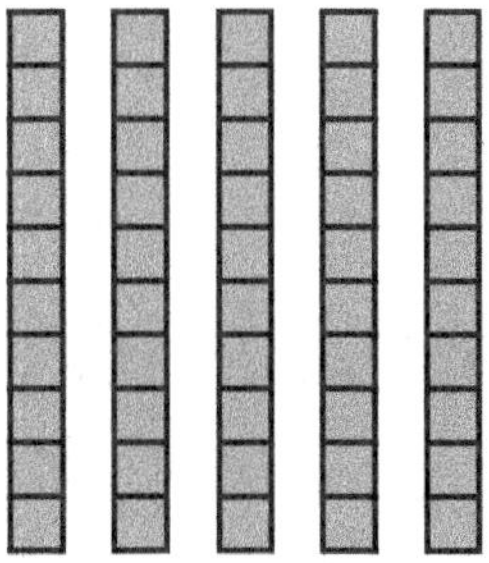

50 − 20 = ______

8.

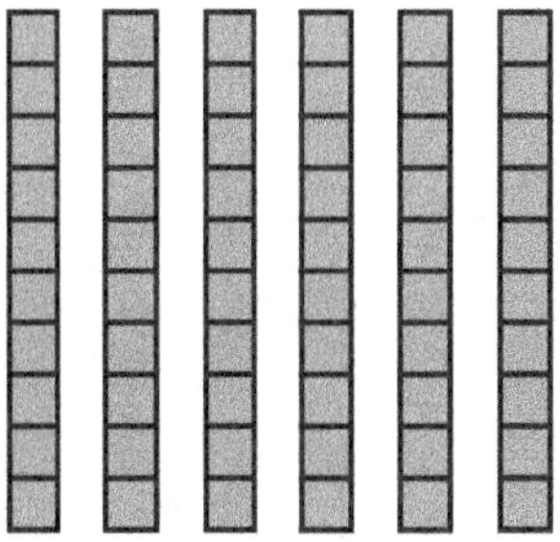

60 − 30 = ______

☐ Draw and cross out tens to subtract.

9.

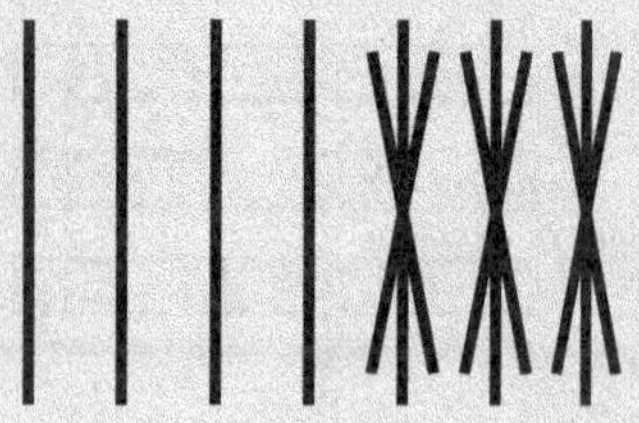

70 − 30 = 40

10.

50 − 30 = ______

11.

60 − 40 = ______

12.

30 − 20 = ______

13.

50 − 10 = ______

14.

60 − 20 = ______

15.

70 − 50 = ______

16.

30 − 30 = ______

Subtract.

17.

8 ones	8	8 tens	80
− 3 ones	− 3	− 3 tens	− 30
□ ones	□	□ tens	□

18.

9 ones	9	9 tens	90
− 6 ones	− 6	− 6 tens	− 60
□ ones	□	□ tens	□

19. $3 - 1 = \square$ $\quad 30 - 10 = \square$

20. $6 - 5 = \square$ $\quad 60 - 50 = \square$

21. $8 - 6 = \square$ $\quad 80 - 60 = \square$

22. $2 - 1 = \square$ $\quad 20 - 10 = \square$

23. $8 - 4 = \square$ $\quad 80 - 40 = \square$

24. $9 - 8 = \square$ $\quad 90 - 80 = \square$

25.
7 − 2 = ____
70 − 20 = ____

26.
6 − 3 = ____
60 − 30 = ____

27.
7 − 4 = ____
70 − 40 = ____

☐ Subtract by using an addition fact.

28.

90 − 20

20 + _____ = 90

So 90 − 20 = _____

29.

50 − 40

40 + _____ = 50

So 50 − 40 = _____

30.

80 − 50

50 + _____ = 80

So 80 − 50 = _____

31.

90 − 40

40 + _____ = 90

So 90 − 40 = _____

32.

70 geese are in a pond.
50 fly away.
How many geese are still in the pond?

33.

Ava says the picture shows 60 − 30.
Explain her mistake.

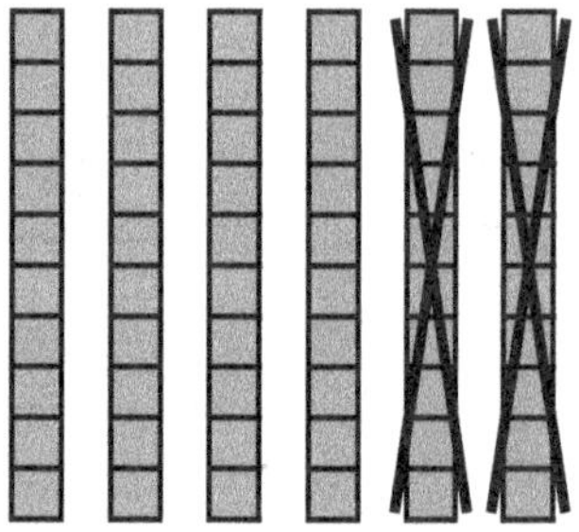

34.

Explain how you know 90 − 70 = 20.

NBT1-30 Adding 10

☐ Circle the next 10 numbers.

☐ Add 10.

1.

1	2	3	4	5	6	7	8	9	10
11	12	13	14	15	16	17	18	19	20

4 + 10 = ____

2.

11	12	13	14	15	16	17	18	19	20
21	22	23	24	25	26	27	28	29	30

19 + 10 = ____

3.

31	32	33	34	35	36	37	38	39	40
41	42	43	44	45	46	47	48	49	50

38 + 10 = ____

4.

81	82	83	84	85	86	87	88	89	90
91	92	93	94	95	96	97	98	99	100

90 + 10 = ____

☐ Add 10 by looking one row below.

5.

1	2	3	4	5	6	7	8	9	10
11	12	13	14	15	16	17	18	19	20

3 + 10 = ____

7 + 10 = ____

9 + 10 = ____

☐ Look one row below to add 10.

6.

1	2	3	4	5	6	7	8	9	10
11	12	13	14	15	16	17	18	19	20
21	22	23	24	25	26	27	28	29	30
31	32	33	34	35	36	37	38	39	40

2 + 10 = ____   8 + 10 = ____   20 + 10 = ____

17 + 10 = ____   25 + 10 = ____   19 + 10 = ____

11 + 10 = ____   23 + 10 = ____   30 + 10 = ____

10 + 17 = ____   10 + 14 = ____   10 + 27 = ____

☐ Add 10.

7. + 10

26 → *36*

8. + 10

13 → ☐

9. + 10

12 → ☐

10. + 10

29 → ☐

11. Which digit stays the same when you add 10?

____

12. Which digit goes up by 1 when you add 10?

____

☐ Add 10 by tracing a tens block.

13. 25 + 10 = 35

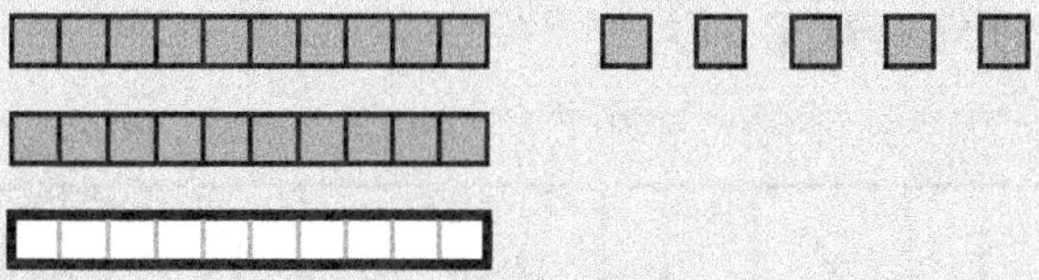

14. 17 + 10 = ____

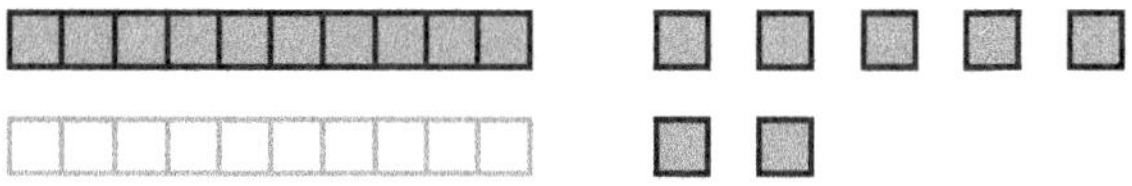

15. 20 + 10 = ____

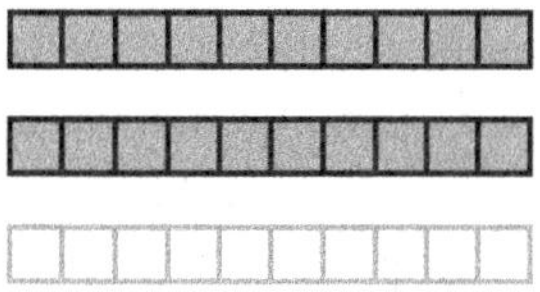

☐ Add 10.

16. 11 + 10 = ____

17. 46 + 10 = ____

18. 39 + 10 = ____

19. 64 + 10 = ____

20. 55 + 10 = ____

21. 77 + 10 = ____

22. 10 + 32 = ____

23. 87 + 10 = ____

24. 63 + 10 = ____

25. 10 + 40 = ____

26. 10 + 71 = ____

27. 86 + 10 = ____

28. Rob writes 78 + 10 = 79. Explain his mistake.

NBT1-31 Subtracting 10

☐ Circle the 10 numbers before ▢.

☐ Subtract 10.

1.

1	2	3	4	5	6	7	8	9	10
11	12	13	14	15	16	17	18	19	20

17 – 10 = _____

2.

11	12	13	14	15	16	17	18	19	20
21	22	23	24	25	26	27	28	29	30

30 – 10 = _____

3.

41	42	43	44	45	46	47	48	49	50
51	52	53	54	55	56	57	58	59	60

52 – 10 = _____

☐ Look one row above to subtract 10.

4.

71	72	73	74	75	76	77	78	79	80
81	82	83	84	85	86	87	88	89	90

82 – 10 = _____

85 – 10 = _____

90 – 10 = _____

5. Which digit stays the same when you subtract 10?

____

6. Which digit is 1 less when you subtract 10?

____

☐ Subtract 10 by crossing out a tens block.

7. 26 − 10 = 16

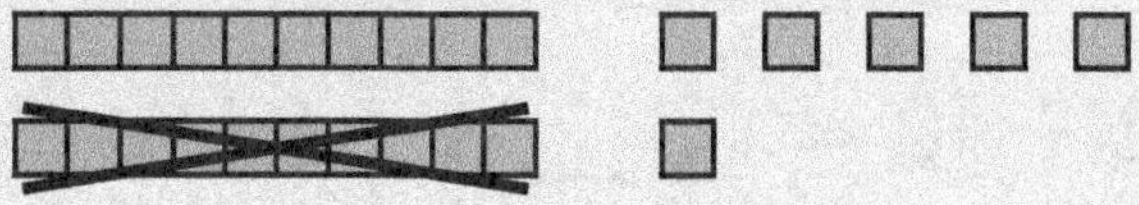

8. 23 − 10 = _____

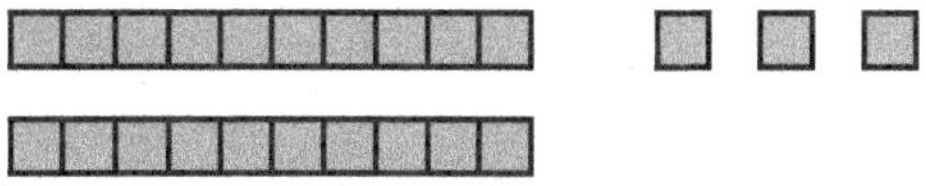

9. 17 − 10 = _____

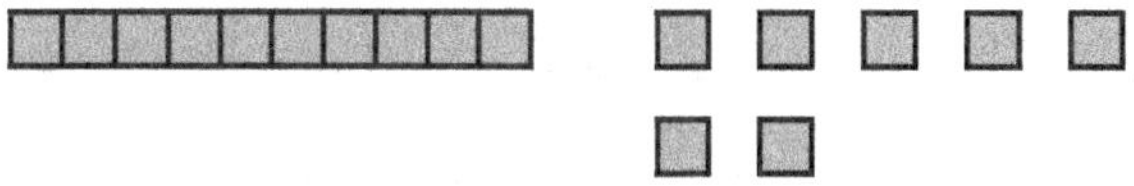

☐ Subtract 10.

10. 18 − 10 = _____

11. 29 − 10 = _____

12. 32 − 10 = _____

13. 76 − 10 = _____

14. 38 − 10 = _____

15. 99 − 10 = _____

16. 35 − 10 = _____

17. 41 − 10 = _____

18. 33 − 10 = _____

19. 62 − 10 = _____

20. 59 − 10 = _____

21. 14 − 10 = _____

22. 28 − 10 = _____

23. 45 − 10 = _____

24. 84 − 10 = _____

NBT1-32 More Adding and Subtracting 10

☐ Write the **tens digit**.

1. 42 + 10 = 5 2

2. 37 + 10 = __ 7

3. 16 + 10 = __ 6

4. 73 + 10 = __ 3

5. 56 + 10 = __ 6

6. 45 + 10 = __ 5

☐ Write the **ones digit**.

7. 24 + 10 = 3 4

8. 62 + 10 = 7 __

9. 28 + 10 = 3 __

10. 40 + 10 = 5 __

11. 66 + 10 = 7 __

12. 53 + 10 = 6 __

☐ Write the missing digit.

13. 13 + 10 = 2 3

14. 87 + 10 = __ 7

15. 68 + 10 = 7 __

16. 47 + 10 = 5 __

17. 32 + 10 = __ 2

18. 75 + 10 = 8 __

☐ Add 10.

19. 43 + 10 = 5 3

20. 7 + 10 = __ __

21. 18 + 10 = __ __

22. 33 + 10 = __ __

23. 25 + 10 = __ __

24. 4 + 10 = __ __

Write the tens digit.

25. 42 − 10 = __ 2

26. 37 − 10 = __ 7

27. 26 − 10 = __ 6

28. 50 − 10 = __ 0

29. 78 − 10 = __ 8

30. 45 − 10 = __ 5

Write the ones digit.

31. 24 − 10 = 1 4

32. 34 − 10 = 2 __

33. 29 − 10 = 1 __

34. 93 − 10 = 8 __

35. 54 − 10 = 4 __

36. 76 − 10 = 6 __

Write the missing digit.

37. 74 − 10 = __ 4

38. 61 − 10 = __ 1

39. 65 − 10 = 5 __

40. 87 − 10 = 7 __

41. 68 − 10 = __ 8

42. 47 − 10 = 3 __

Subtract 10.

43. 41 − 10 = 3 1

44. 17 − 10 = __ 7

45. 38 − 10 = __ __

46. 28 − 10 = __ __

47. 80 − 10 = __ __

48. 56 − 10 = __ __

NBTI-33 Ten More or Ten Less

☐ Trace a tens block to find 10 more.

1.

23
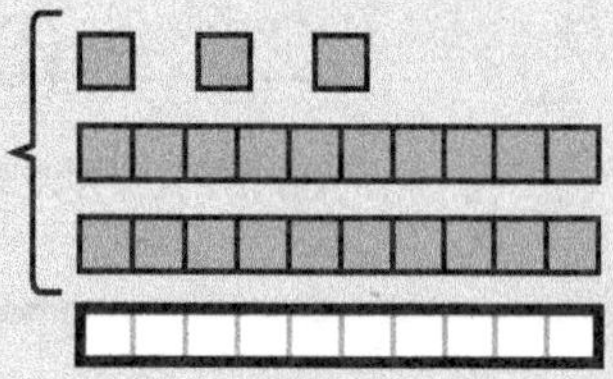

10 more than 23 is *33*.

2.

25
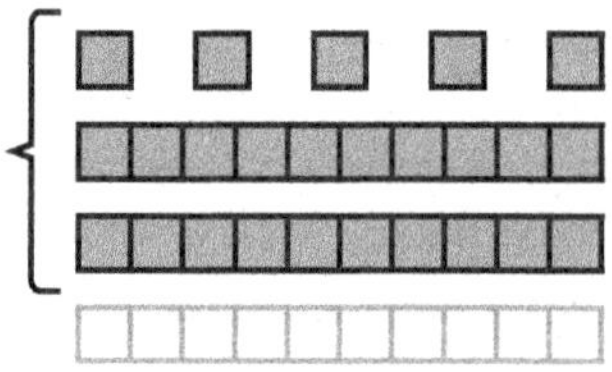

10 more than 25 is ______.

3.

24
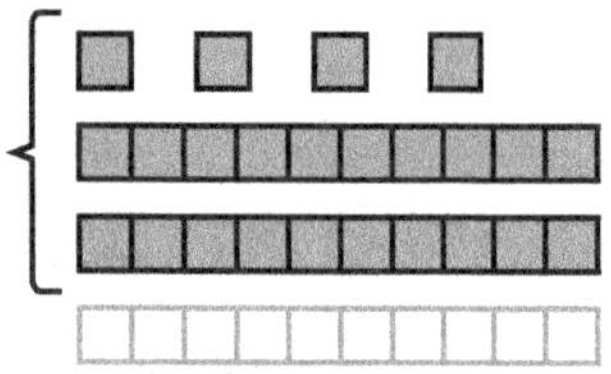

10 more than 24 is ______.

4.

14
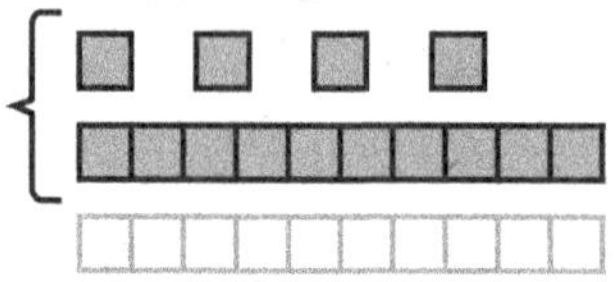

10 more than 14 is ______.

5.

31
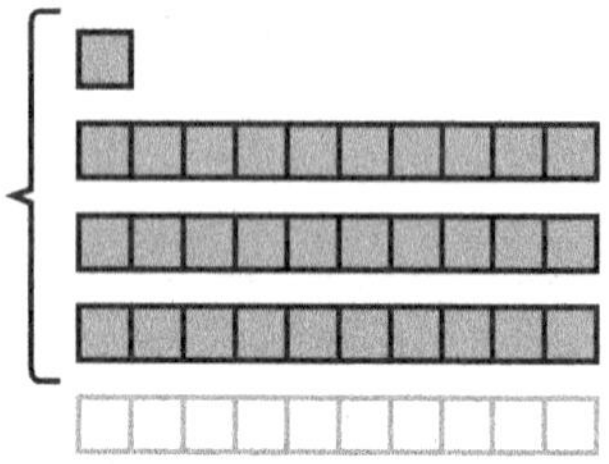

10 more than 31 is ______.

6.

35
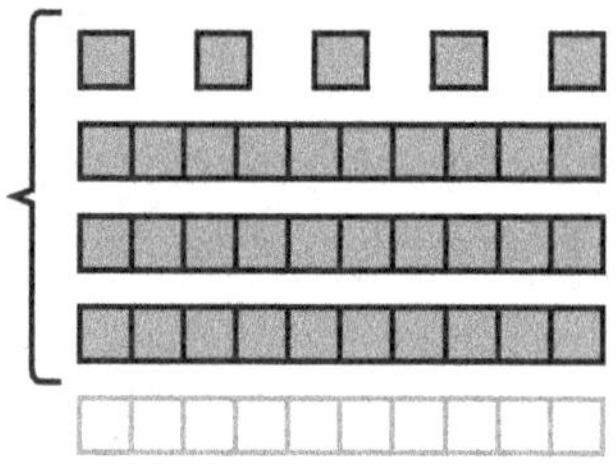

10 more than 35 is ______.

☐ Find the answer.

☐ Write the addition fact you used.

7. 10 more than 42 is *52*.

I used *42 + 10 = 52*.

8. 10 more than 63 is ______.

I used ______________.

9. 10 more than 81 is ______.

I used ______________.

10. 10 more than 28 is ______.

I used ______________.

☐ Cross out a tens block to find 10 less.

11.

32
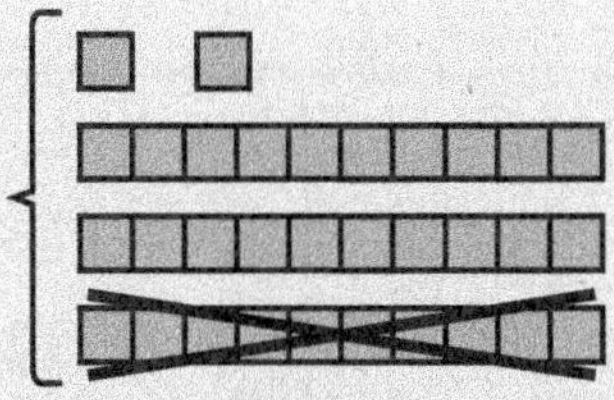

10 less than 32 is __22__.

12.

25
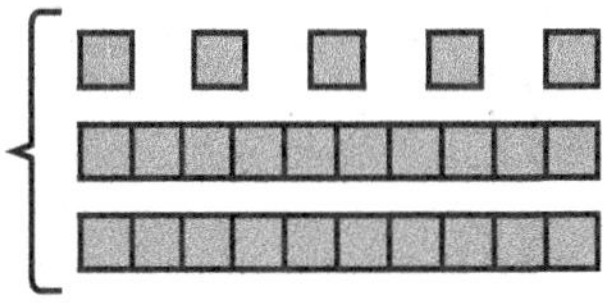

10 less than 25 is _____.

13.

23
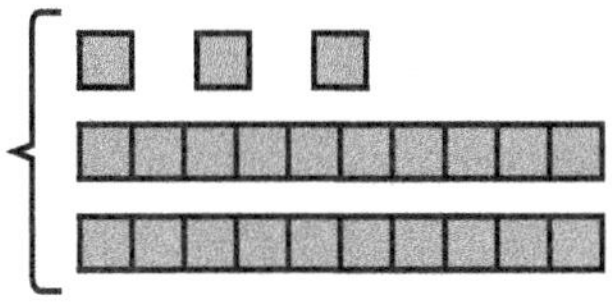

10 less than 23 is _____.

14.

34
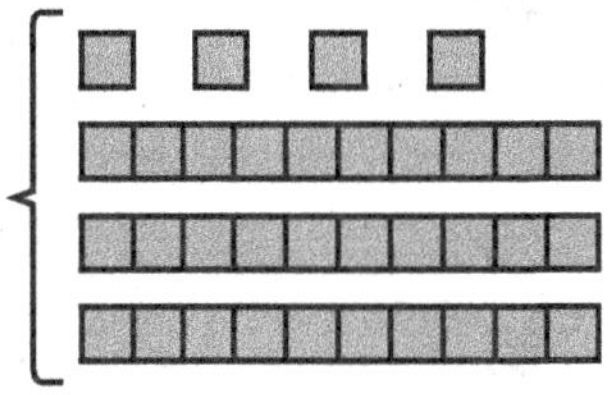

10 less than 34 is _____.

☐ Find the answer.

☐ Write the subtraction fact you used.

15. 10 less than 63 is _______.

I used _________________.

16. 10 less than 42 is _______.

I used _________________.

☐ Write the numbers that are 10 less and 10 more.

17. __17__ 27 __37__

18. _____ 61 _____

19. _____ 58 _____

20. Jenny says 57 is 10 more than 47. Is she correct? Explain.

21. What number is 10 less than 10? How do you know?

NBTI-34 Counting On by Tens from Any Number

☐ Count on by tens.

1.

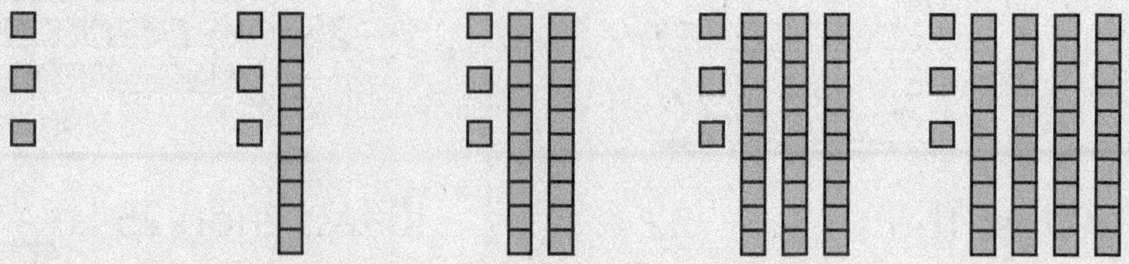

3   13   23   33   43

2.

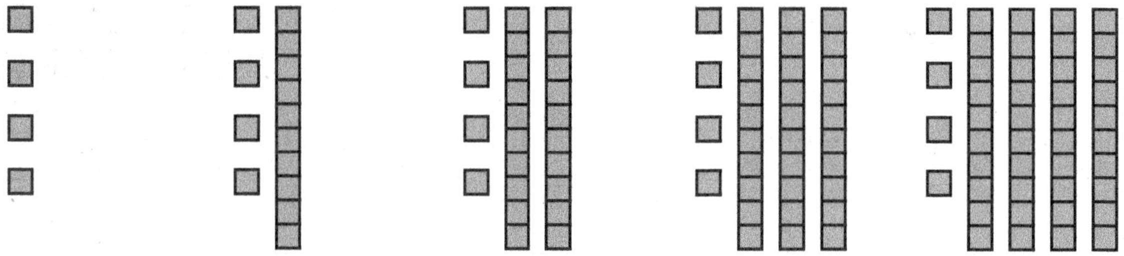

____   ____   ____   ____   ____

3.

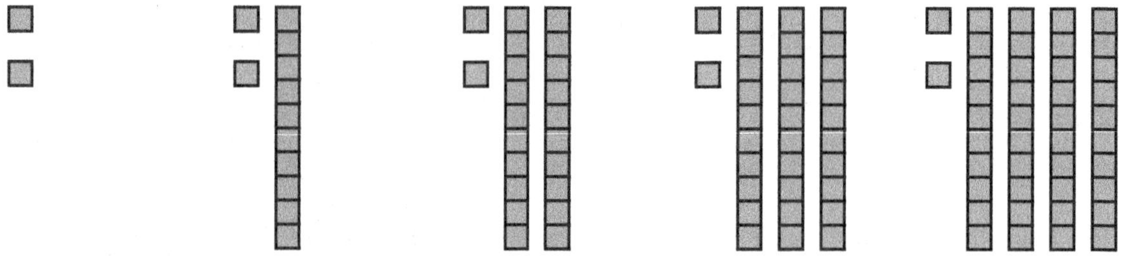

____   ____   ____   ____   ____

4.

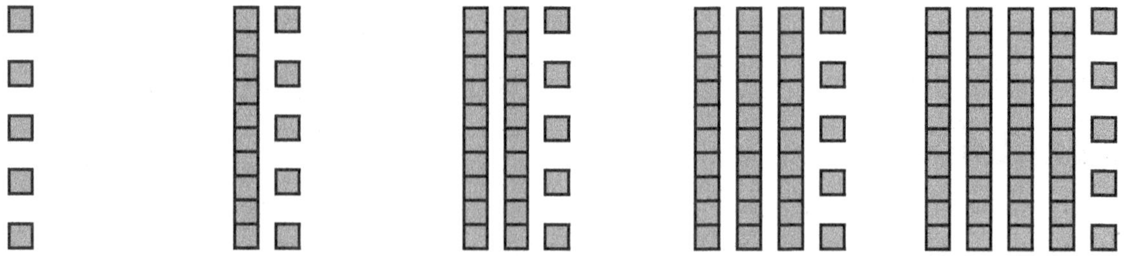

____   ____   ____   ____   ____

☐ Count on by tens.

5.

7, 17, 27, ____, ____, ____

8, ____, ____, ____, ____, ____

32, 42, ____, ____, ____, ____

25, ____, ____, ____, ____, ____

13, ____, ____, ____, ____, ____

6.

There are 13 fish in a pond.
Mary puts 10 more fish in the pond every day.
How many fish are in the pond on Day 3?

13 ____ ____ ____

Day 1 Day 2 Day 3

7.

Alexa counts on by tens.

2, 12, 22, 32, 42, 53

What mistake does she make?

NBTI-35 Counting On by Tens to Add

☐ Circle the number with no ones.

☐ How many tens are in that number?

1. 35 + (20) 2 tens	2. (30) + 42 ____ tens	3. 75 + 20 ____ tens
4. 40 + 15 ____ tens	5. 23 + 50 ____ tens	6. 61 + 30 ____ tens
7. 16 + 60 ____ tens	8. 58 + 40 ____ tens	9. 70 + 27 ____ tens

☐ Start with the number that has ones.

☐ Count on by tens to add.

10. 42 + 30 (42) $\xrightarrow{+10}$ 52 $\xrightarrow{+10}$ 62 $\xrightarrow{+10}$ 72

11. 30 + 37 (37) ____ ____ ____

12. 35 + 20 ( ) ____ ____

13. 64 + 30 ( ) ____ ____ ____

Peter uses his fingers to add 54 + 30.

3 tens, so 3 fingers up

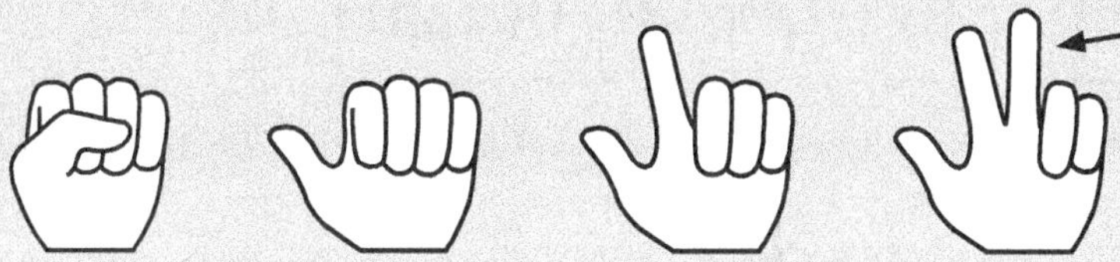

54 64 74 84

54 + 30 = 84

☐ Circle the number that has ones.

☐ Count on by tens to add.

14. 20 + (55) = 75

15. 32 + 10 = ____

16. 45 + 30 = ____

17. 10 + 72 = ____

18. 63 + 10 = ____

19. 20 + 71 = ____

20. 56 + 30 = ____

21. 48 + 40 = ____

22. 80 + 17 = ____

☐ Count on by tens to add.

23. 23 + 40 = 63

24. 34 + 20 = ____

25. 15 + 30 = ____

26. 38 + 20 = ____

27. 27 + 50 = ____

28. 39 + 60 = ____

29. 10 + 82 = ____

30. 30 + 44 = ____

31. 50 + 41 = ____

NBTI-36 Hundreds Chart Pieces

1	2	3	4	5	6	7	8	9	10
11	12	13	14	15	16	17	18	19	20

The boxes are pieces from a hundreds chart.

☐ Find the missing numbers.

1. Add 1.

3	

17	

19	

6	

2. Add 10.

23	30	6	24	31	29

3. Add 1 or 10.

45	

50

41	

52

53

56	

4. Add 1 and 10 in the correct order.

71	

77	

74	

86	

82	

☐ Find the missing numbers.

5. Subtract 1 or 10.

	48

57

	43

43

41

	40

6. Subtract 1 and 10 in the correct order.

92, 89, 77, 85, 90

7. Add or subtract 1 or 10.

26, 37, 39, 39, 39, 39

8. Add or subtract or both.

18, 24, 17, 38, 15

35, 19, 46, 18, 47

☐ Write the missing numbers on the hundreds chart piece.

9.

33		35		37
	44	45	46	47
53	54		56	

10.

11	12	13	
21		23	24
	32	33	

11.

61	62	
71		73
		83
91	92	

12.

64		66
	75	
	85	
94		96

13.

	49
58	
	69
78	

14.

46				50
	57		59	
		68		
	77		79	
86				90

15.

	57	58	59	60
66				
	77			
		88		
			99	100

NBTI-37 Adding Tens and Ones

☐ Write the number as adding tens and ones.

1.

32 = 10 + 10 + 10 + 1 + 1

2.

23 = ____ + ____ + ____ + ____ + ____

3.

14 = ____ + ____ + ____ + ____ + ____

4.

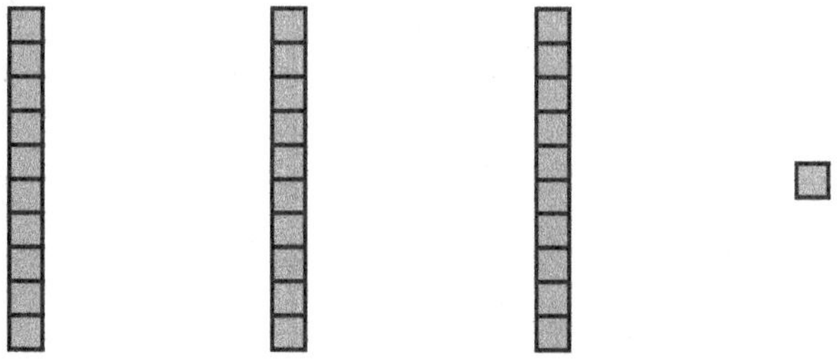

31 = ____ + ____ + ____ + ____

☐ Write the number as adding tens and ones.

5. 22 = 10 + 10 + 1 + 1

6. 33 = ______

7. 41 = ______

8. 25 = ______

9. 62 = ______

☐ Add.

10.

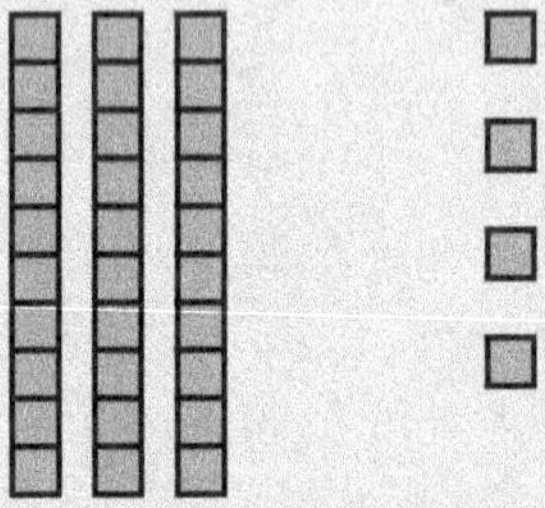

30 + 4 = 34

11.

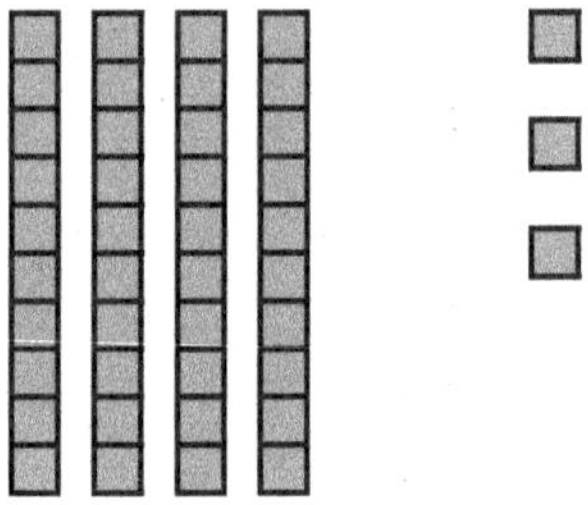

40 + 3 = ______

12.

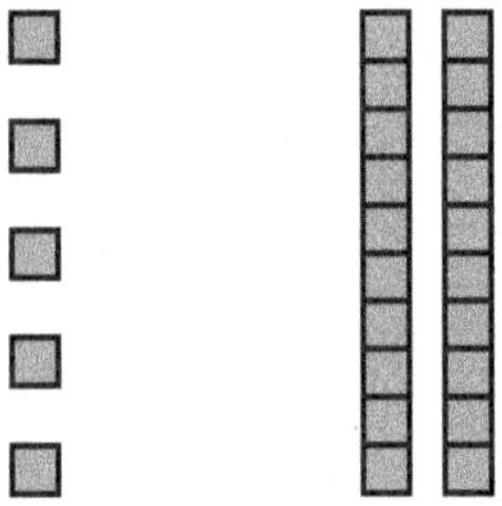

5 + 20 = ______

13.

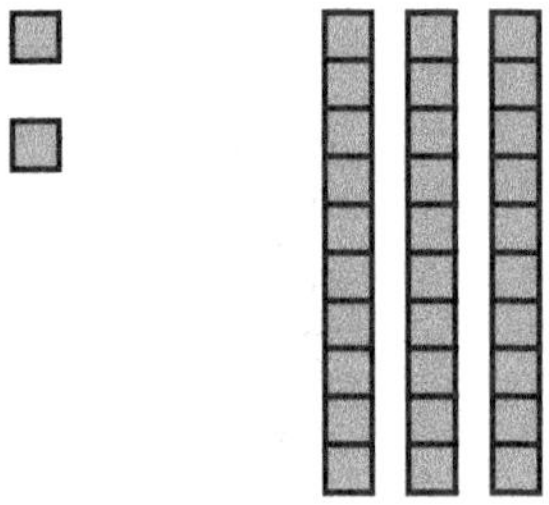

2 + 30 = ______

☐ Add.

14. $40 + 5 =$ <u>45</u>

15. $6 + 20 =$ ____

16. $70 + 1 =$ ____

17. $8 + 60 =$ ____

18. $70 + 7 =$ ____

19. $4 + 50 =$ ____

20. $30 + 8 =$ ____

21. $9 + 10 =$ ____

22. $6 + 80 =$ ____

23. $7 + 90 =$ ____

24. $9 + 70 =$ ____

25. $70 + 9 =$ ____

You can write $24 = 20 + 4$.

☐ Write the number in the same way.

26. $35 =$ <u>30 + 5</u>

27. $47 =$ ______

28. $63 =$ ______

29. $81 =$ ______

30. $56 =$ ______

31. $92 =$ ______

32. $23 =$ ______

33. $87 =$ ______

34. $79 =$ ______

35. Sandy writes $37 = 30 + 70$.
What mistake does she make?

NBT1-38 Adding a 1-Digit Number to a 2-Digit Number

☐ How many tens and ones?

☐ Add.

1.

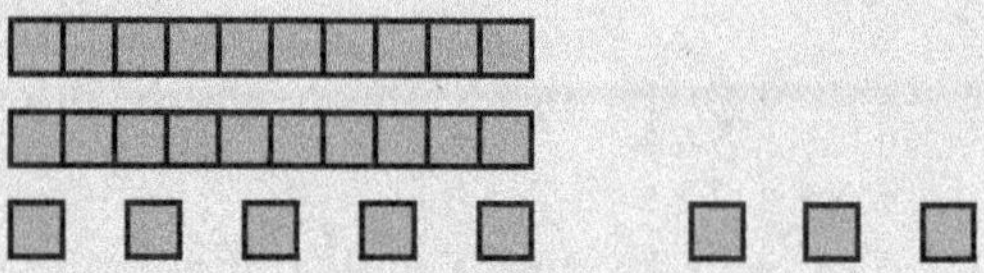

= __2__ tens + __8__ ones

25 + 3 = __28__

2.

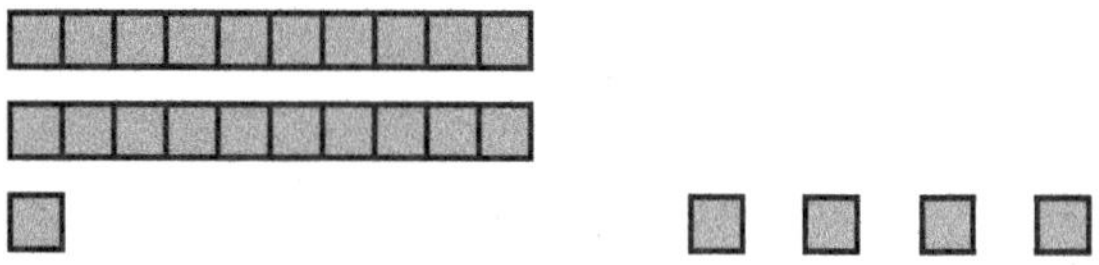

= ____ tens + ____ ones

21 + 4 = ____

3.

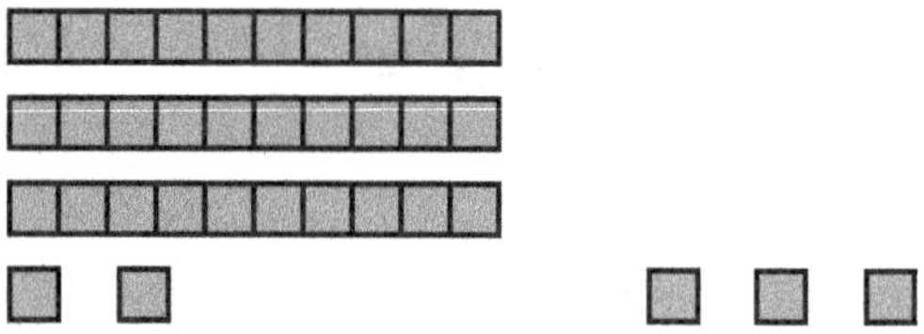

= ____ tens + ____ ones

32 + 3 = ____

4. BONUS

= ____ ten + ____ ones

5 + 12 = ____

☐ Add.

5. $37 + 2 = $ ____

6. $63 + 1 = $ ____

7. $72 + 3 = $ ____

8. $14 + 5 = $ ____

9. $52 + 2 = $ ____

10. $33 + 6 = $ ____

11. $8 + 91 = $ ____

12. $45 + 4 = $ ____

13. $26 + 3 = $ ____

14. ____ $= 4 + 35$

15. ____ $= 5 + 43$

16. $82 + 6 = $ ____

17. $72 + 7 = $ ____

18. ____ $= 11 + 6$

19. ____ $= 22 + 5$

20.

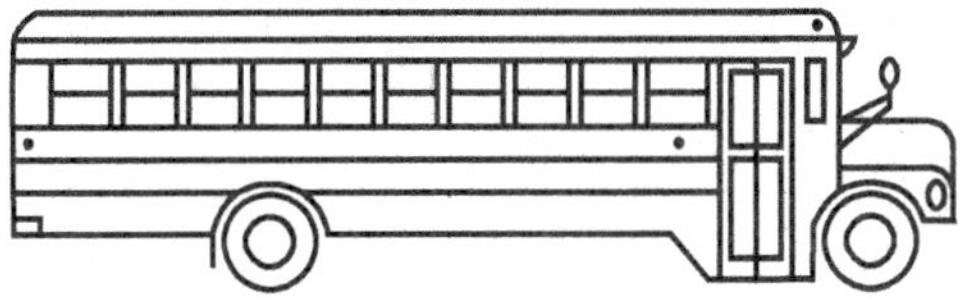

24 children are on a bus.
3 more get on.
How many children are on the bus?

21.

Glen has 5 marbles.
Jay has 33 marbles.
How many do they have altogether?

NBTI-39 Adding by Counting by Tens Then Ones

☐ Add by counting on by tens then ones.

1.

35 + 23

35 | 23

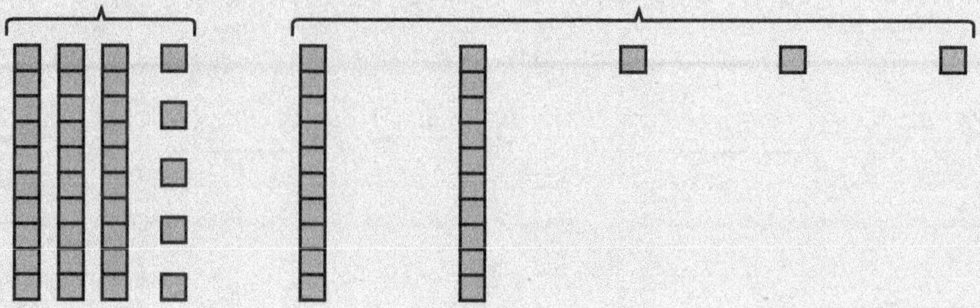

35   45   55   56   57   58

2.

26 + 32

26 | 32

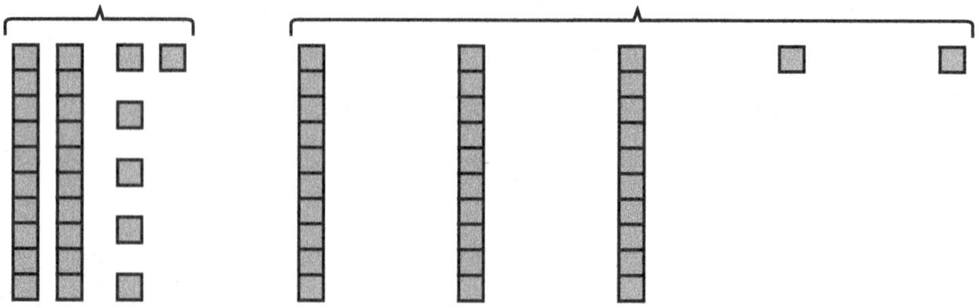

26 ____ ____ ____ ____ ____

3.

32 + 41

32 | 41

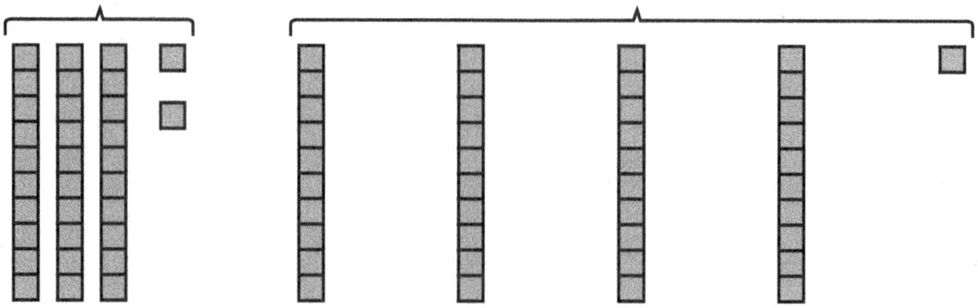

32 ____ ____ ____ ____ ____

☐ Write the second number as adding tens and ones.

☐ Add by counting on by tens then ones.

4.

25 + 32

25 + 10 + 10 + 10 + 1 + 1

25 35 45 55 56 57

So 25 + 32 = 57

5.

42 + 23

42 + ____ + ____ + ____ + ____ + ____

42 ____ ____ ____ ____ ____

So 42 + 23 = ____

6.

51 + 21

51 + ____ + ____ + ____

51 ____ ____ ____

So 51 + 21 = ____

7.

67 + 12

67 + ____ + ____ + ____

67 ____ ____ ____

So 67 + 12 = ____

☐ Add by counting on by tens then ones.

8. 55 + 23 = ____

9. 61 + 17 = ____

10. 46 + 33 = ____

NBTI-40 Regrouping 10 to Add

☐ Circle ones blocks to make 10.

☐ Add.

1.

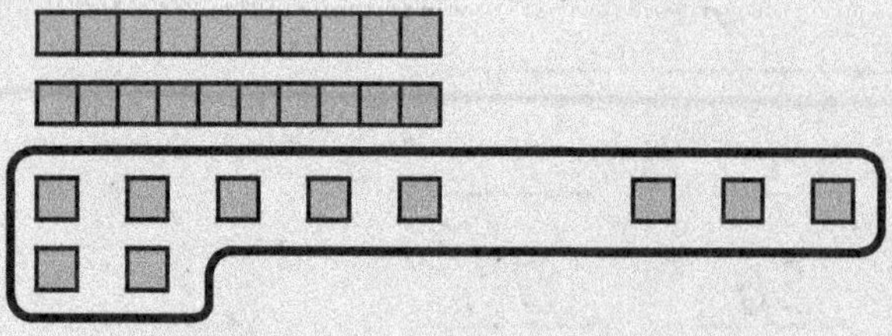

27 + 3 = 30

2.

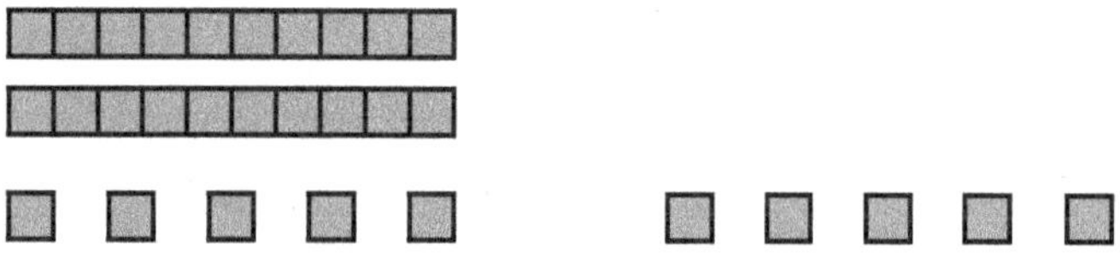

25 + 5 = ____

3.

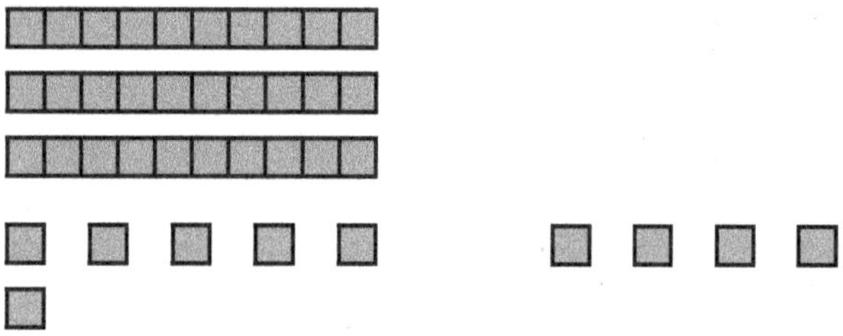

36 + 4 = ____

4.

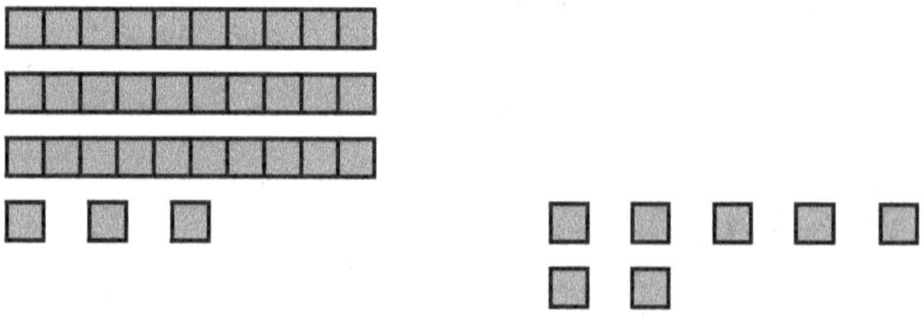

33 + 7 = ____

☐ Circle ones blocks to make 10.

☐ Add.

5.

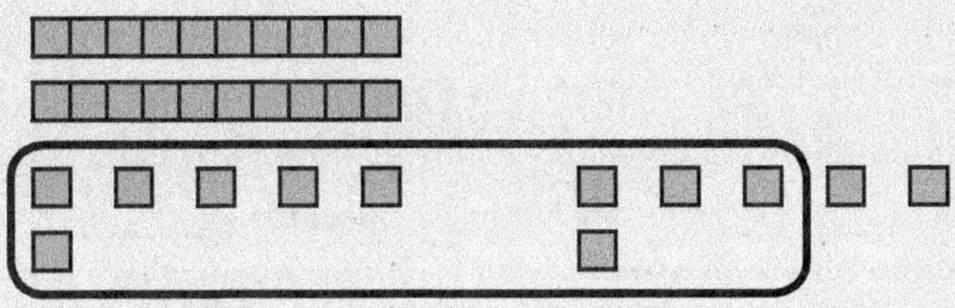

26 + 6 = 32

6.

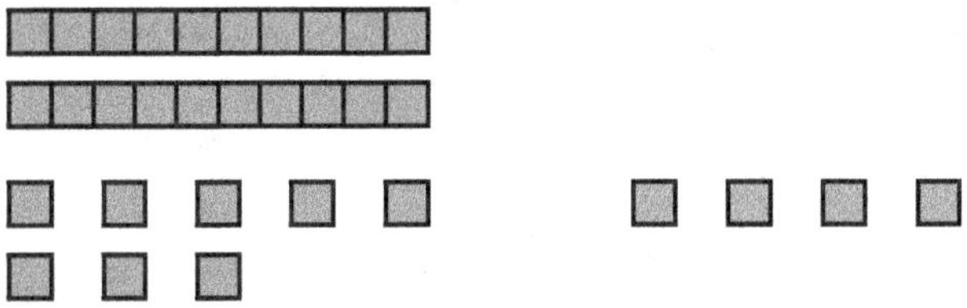

28 + 4 = ______

7.

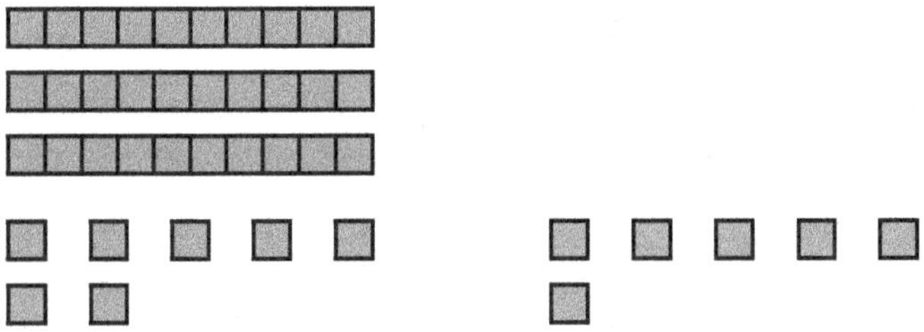

37 + 6 = ______

8.

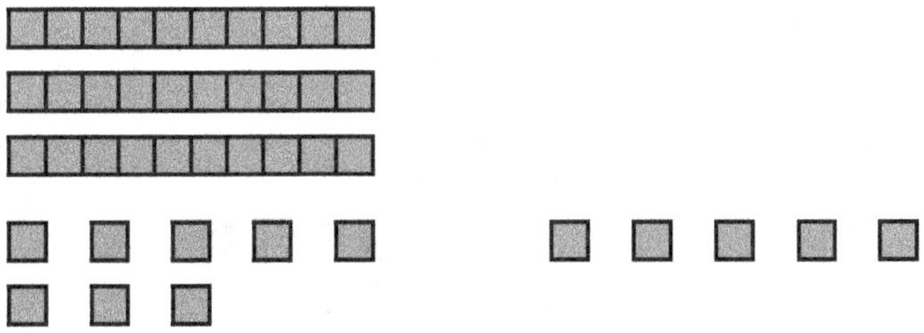

38 + 5 = ______

☐ Fill in the blank.

9. 7 + __3__ = 10

10. 2 + ____ = 10

11. 6 + ____ = 10

12. 1 + ____ = 10

13. 5 + ____ = 10

14. 8 + ____ = 10

☐ Add to make the next 10.

15. 7 + ____ = 10 47 + ____ = 50 87 + ____ = 90

16. 6 + ____ = 10 26 + ____ = 30 56 + ____ = 60

☐ Fill in the blank to make 10 with the first number.

17. 8 + 5
8 + __2__ + 3
10

18. 6 + 9
6 + ___ + 5
10

19. 3 + 9
3 + ___ + 2
10

20. 7 + 8
7 + ___ + 5
10

21. 9 + 4
9 + ___ + 3
10

22. 5 + 7
5 + ___ + 2
10

23. 4 + 8
4 + ___ + 2
10

24. 9 + 9
9 + ___ + 8
10

☐ When you make the next 10, how many are left?

25. 27 + 7 → 27 + 3 + 4

26. 28 + 5 → 28 + ___ + ___

27. 37 + 4 → 37 + ___ + ___

28. 64 + 8 → 64 + ___ + ___

29. 83 + 8 → 83 + ___ + ___

30. 29 + 5 → 29 + ___ + ___

31. 18 + 9 → 18 + ___ + ___

32. 26 + 6 → 26 + ___ + ___

33. 57 + 7 → 57 + ___ + ___

☐ Add.

34. 38 + 5
= 38 + 2 + 3
= 40 + 3
= 43

35. 29 + 6
= 29 + ___ + ___
= ______ + ___
= ______

36. 49 + 3
= 49 + ___ + ___
= ______ + ___
= ______

37. 67 + 8
= 67 + ___ + ___
= ______ + ___
= ______

38. 35 + 9
= 35 + ___ + ___
= ______ + ___
= ______

39. 46 + 6
= 46 + ___ + ___
= ______ + ___
= ______

NBTI-41 Using Tens and Ones to Add

☐ Fill in the tens and ones.

☐ Add.

1.

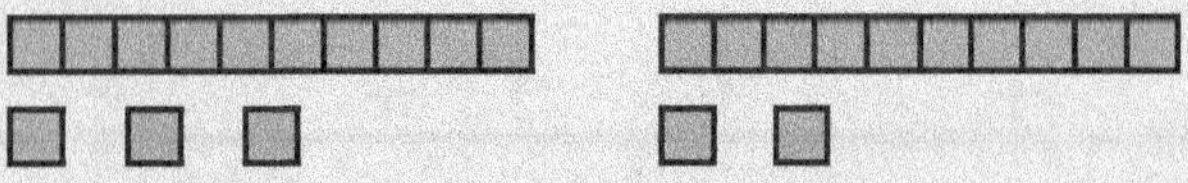

= __2__ tens + __5__ ones

13 + 12 = __25__

2.

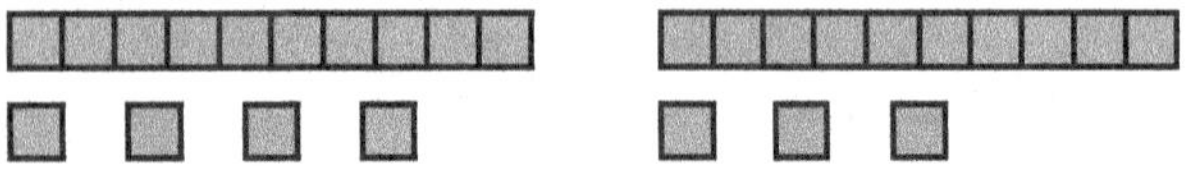

= ____ tens + ____ ones

14 + 13 = ____

3.

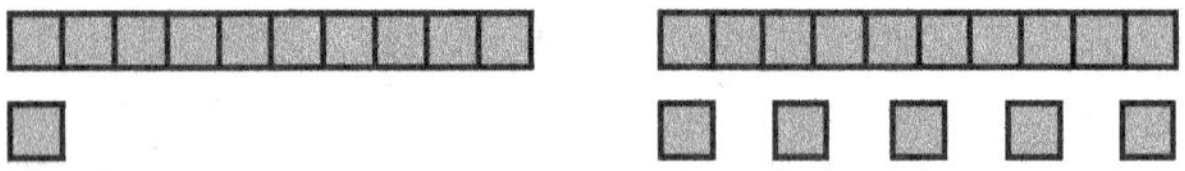

= ____ tens + ____ ones

11 + 15 = ____

☐ Draw the blocks and add.

4.

= ____ tens + ____ ones

12 + 12 = ____

☐ Make your own question.

5.

= ____ tens + ____ ones

= ____

☐ Add by writing the tens and ones.

6.

23	=	20	+	3
+ 34	=	30	+	4
57	←	50	+	7

7.

34	=	30	+	4
+ 15	=	10	+	5
☐	←	40	+	9

8.

27	=	20	+	☐
+ 22	=	20	+	☐
☐	←	40	+	☐

9.

35	=	☐	+	☐
+ 42	=	☐	+	☐
☐	←	☐	+	☐

10.

34	=	☐	+	☐
+ 54	=	☐	+	☐
☐	←	☐	+	☐

11.

26	=	☐	+	☐
+ 33	=	☐	+	☐
☐	←	☐	+	☐

12.

22	=	☐	+	☐
14	=	☐	+	☐
+ 21	=	☐	+	☐
☐	←	☐	+	☐

13.

11	=	☐	+	☐
22	=	☐	+	☐
+ 33	=	☐	+	☐
☐	←	☐	+	☐

☐ Add by using a tens and ones chart.

14.

  35
+ 32
= 67

Tens	Ones
3	5
3	2
6	7

15.

  24
+ 41

Tens	Ones
2	4
4	1

16.

  46
+ 31

Tens	Ones

17.

  43
+ 23

Tens	Ones

18.

  27
+ 21
+ 51

Tens	Ones

19.

  31
+ 42
+ 14

Tens	Ones

20.

37	63	25	31	54	23
+ 22	+ 16	+ 34	+ 62	+ 34	+ 43

NBTI-42 Regrouping (Advanced)

- [ ] Group 10 ones blocks.
- [ ] Write the tens and ones.
- [ ] Add.

1.

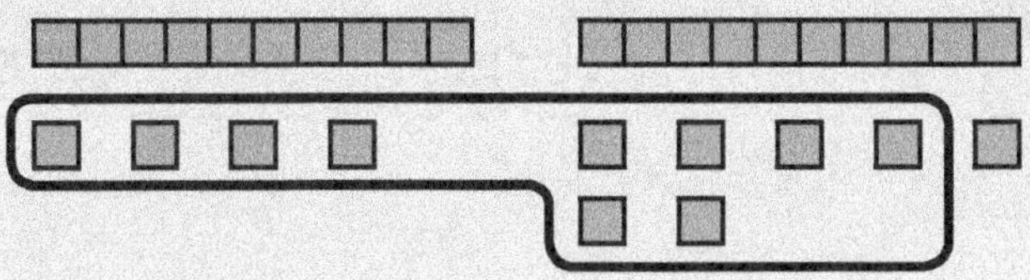

= 3 tens + 1 one

14 + 17 = 31

2.

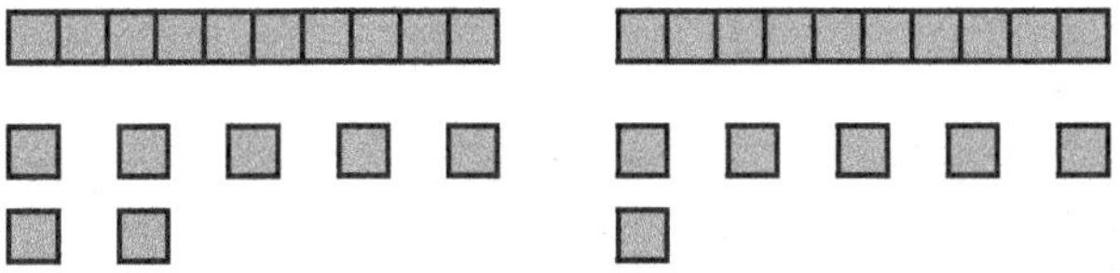

= _____ tens + _____ ones

17 + 16 = _____

3.

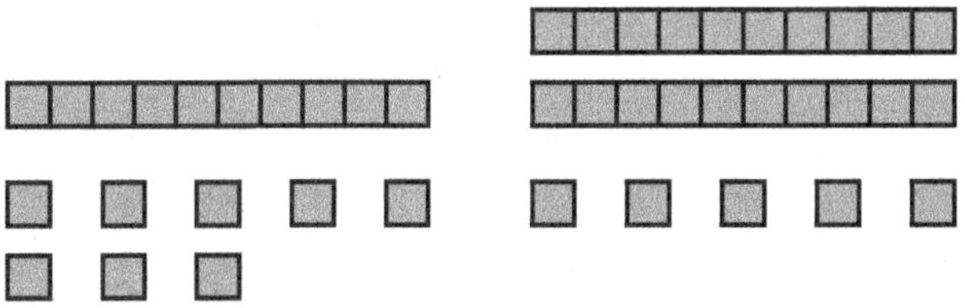

= _____ tens + _____ ones

18 + 25 = _____

4.

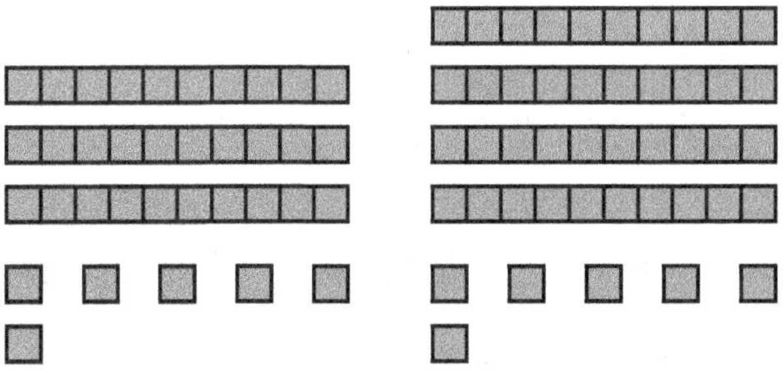

= _____ tens + _____ ones

36 + 46 = _____

- [ ] Group 10 ones blocks.
- [ ] Write the tens and ones.
- [ ] Add.

5.

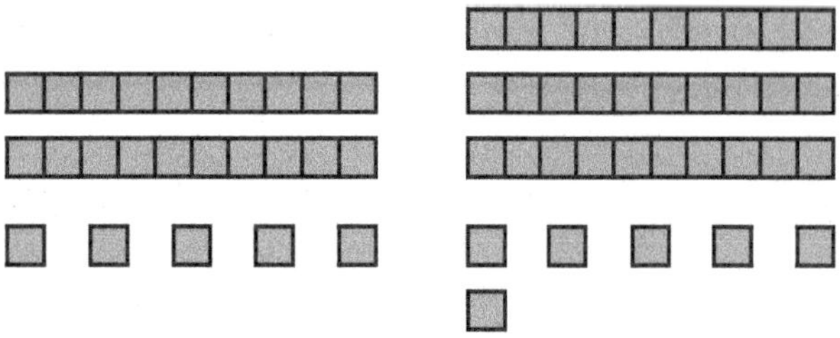

= _____ tens + _____ ones

25 + 36 = _____

6.

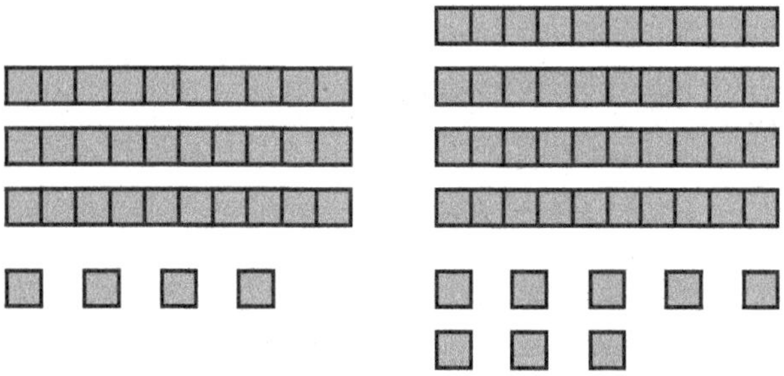

= _____ tens + _____ ones

34 + 48 = _____

7.

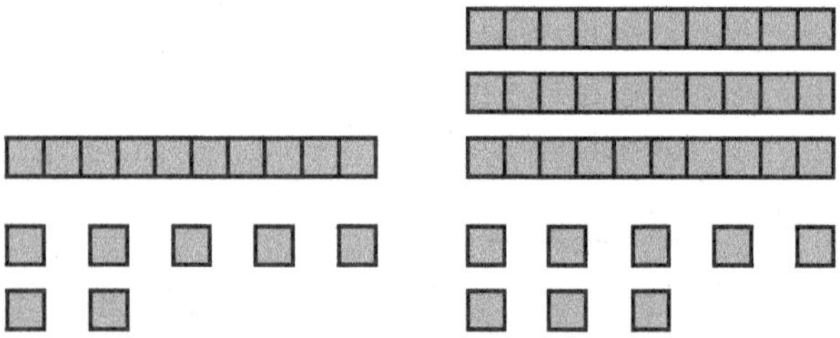

= _____ tens + _____ ones

17 + 38 = _____

8.

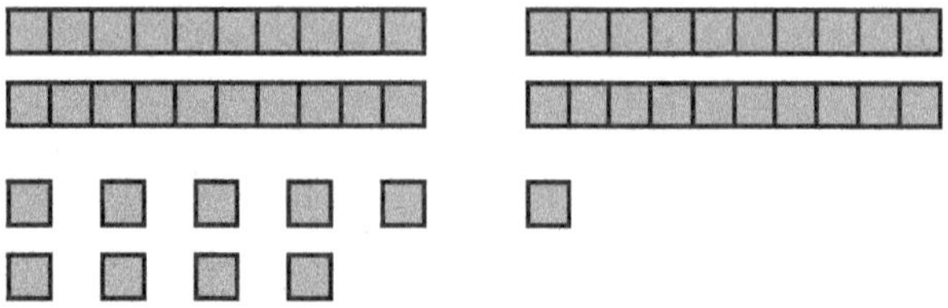

= _____ tens + _____ ones

29 + 21 = _____

NBTI-43 Review

☐ Count on by tens.

1.

36, __46__, __56__, ____, ____, ____

25, ____, ____, ____, ____, ____

☐ Count on by tens then ones to add.

2.

23 + 32 = ____

23, ____, ____, ____, ____, ____

3.

37 + 12 = ____

37, ____, ____, ____

4.

62 + 23 = ____

62, ____, ____, ____, ____, ____

☐ Add by writing tens and ones.

5.

21	=	20	+	☐
+ 34	=	30	+	☐
☐	←	☐	+	☐

6.

46	=	☐	+	☐
+ 52	=	☐	+	☐
☐	←	☐	+	☐

☐ Which addition gives the same number? Write a ✓.

7. 30 + 7

_____ 10 + 7

_____ 10 + 10 + 10 + 7

_____ 10 + 10 + 10 + 10 + 7

8. 30 + 2

_____ 10 + 1 + 1

_____ 10 + 10 + 10 + 1 + 1

_____ 10 + 10 + 10 + 10 + 1 + 1

☐ Subtract by using an addition fact.

9. 70 − 20

20 + _____ = 70

So 70 − 20 = _____

10. 80 − 60

60 + _____ = 80

So 80 − 60 = _____

11. Which is greater? 32 + 10 or 11 + 30

12. Jim has 37 stickers. Carlos has 21 stickers.
How many do they have altogether?

13. Sandy has 40 cherries. She eats 10 of them.
How many does she have now?

OAI-54 Number Sentences (I)

Some balls are hidden in a box.

The same number of balls is on each table.

☐ Draw the hidden balls.

☐ How many balls are hidden?

1.

4 = 4

2.

3 = ☐

3.

☐ = 5

4.

2 = ☐

☐ Draw the hidden balls.

☐ Write the number of balls that are hidden.

5.

2 + 3 = 5

6.

1 + 3 = ☐

7.

2 + 2 = ☐

8.

☐ = 2 + 1

☐ Draw the hidden balls.

☐ Write the number of balls that are hidden.

9.

○○○ + ○ = [○○○○]

3 + 1 = [4]

10.

○○ + ○○ = [ ]

2 + 2 = [ ]

11.

1 + 4 = [○○○○○]

1 + 4 = [ ]

12.

2 + 1 = [ ]

2 + 1 = [ ]

13.

1 + 3 = [ ]

1 + 3 = [ ]

14.

3 + 2 = [ ]

3 + 2 = [ ]

15.

1 + 1 = [ ]

1 + 1 = [ ]

OAI-55 Number Sentences (2)

☐ Draw the hidden balls.

☐ Write the number of balls that are hidden.

1.

2 + [3] = 5

2.

2 + □ = 4

3.

4 = 1 + □

4.

□ + 4 = 6

☐ Draw the hidden balls.

☐ Write the number of balls that are hidden.

5.

○○ + [○○○] = ○○○○○

2 + [3] = 5

6.

○○ + [   ] = ○○○

2 + [   ] = 3

7.

4 + [   ] = 6

4 + [   ] = 6

8.

2 + [   ] = 5

2 + [   ] = 5

9.

3 + [   ] = 6

3 + [   ] = 6

10.

3 + [   ] = 7

3 + [   ] = 7

11.

5 + [   ] = 8

5 + [   ] = 8

OAI-56 Adding and How Many More

- [ ] Draw 3 more circles.
- [ ] Write the addition.

1. 3 more than 2

2 + 3

2. 3 more than 4

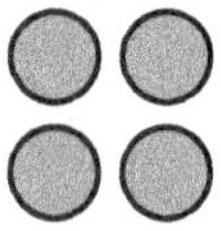

4 + ____

3. 3 more than 6

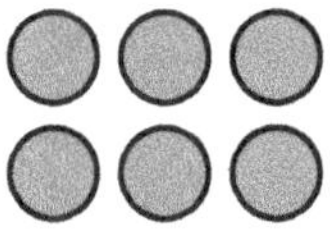

____ + ____

4. 3 more than 3

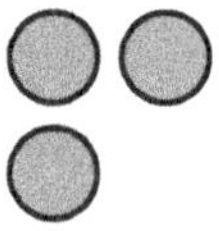

____ + ____

5. 3 more than 5

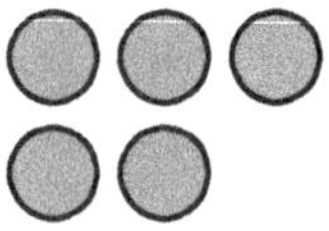

____ + ____

6. 3 more than 1

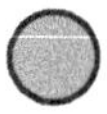

____ + ____

7. 3 more than 7

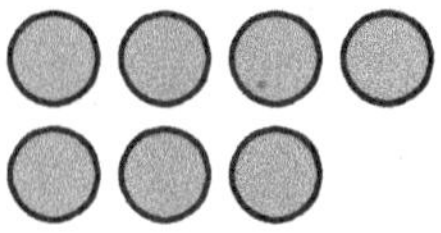

____ + ____

8. 3 more than 8

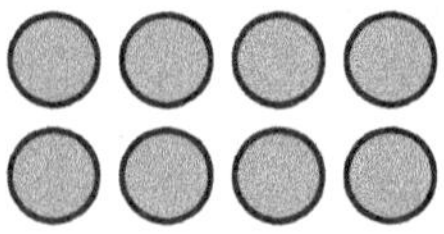

____ + ____

☐ Draw more circles.

☐ Write the addition.

9. 2 more than 3

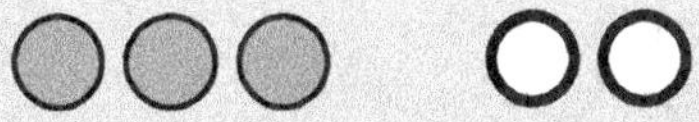

3 + 2

10. I more than 6

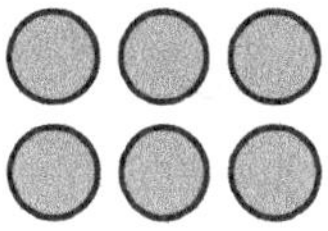

6 + ____

11. 3 more than 5

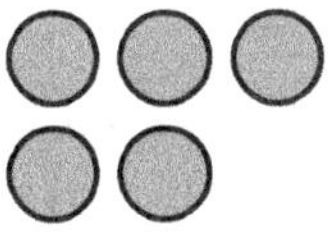

____ + ____

12. 2 more than 7

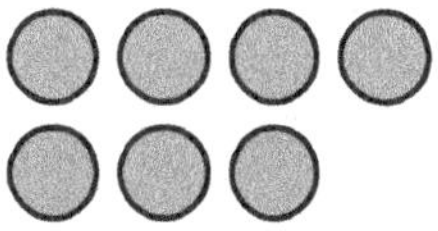

____ + ____

13. 5 more than 4

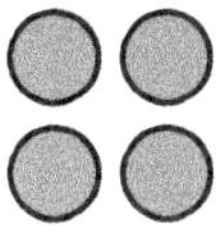

____ + ____

14. 3 more than 8

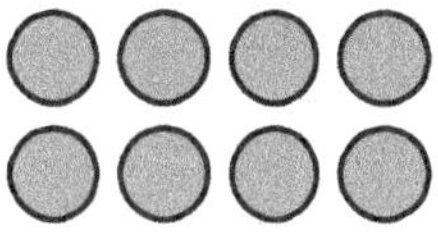

____ + ____

15. 2 more than 6

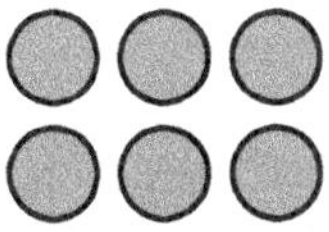

____ + ____

16. 4 more than 3

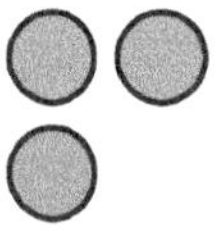

____ + ____

☐ Draw more circles.

☐ Write the **addition sentence**.

17. 5 more than 2

2 + 5 = 7

18. 2 more than 6

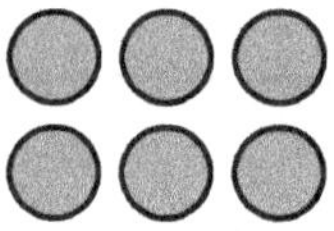

_____ + _____ = _____

19. 4 more than 5

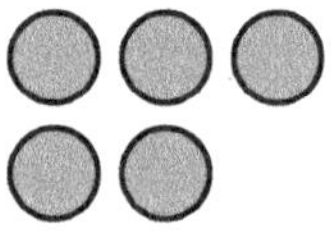

_____ + _____ = _____

20. 6 more than 2

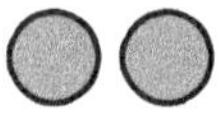

_____ + _____ = _____

☐ Write an addition sentence to find the answer.

21. 3 more than 5

5 + 3 = 8

3 more than 5 is 8.

22. 4 more than 6

_____ + _____ = _____

4 more than 6 is _____.

23. 7 more than 10

_____ + _____ = _____

7 more than 10 is _____.

24. 2 more than 16

_____ + _____ = _____

2 more than 16 is _____.

OAI-57 Finding an Addend by Counting On

☐ Draw circles to find the missing number.

1. $3 + \boxed{2} = 5$

2. $2 + \square = 6$

3. $1 + \square = 5$

4. $3 + \square = 6$

5. $4 + \square = 7$

6. $5 + \square = 6$

7. $\square + 2 = 5$

8. $\square + 3 = 5$

9. $\square + 2 = 4$

10. $\square + 1 = 3$

Raj **counts on** to find the missing number in the number sentence.

$$4 + \square = 6$$

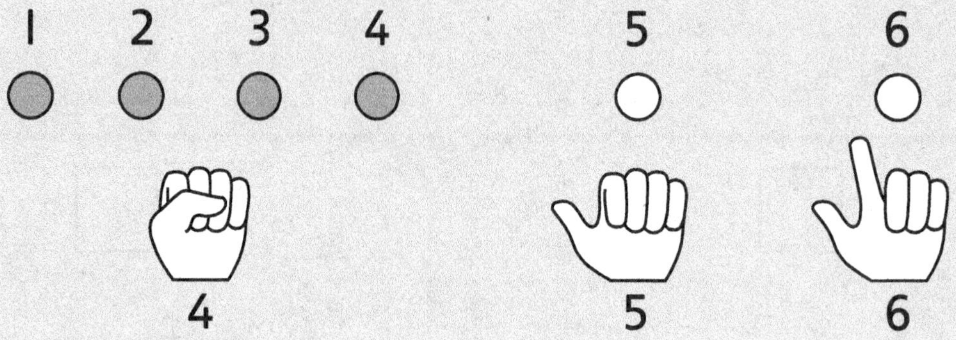

He says "4" with his fist closed.
He counts up from 4 until he reaches 6.
The answer is the number of fingers up.

$$4 + \boxed{2} = 6$$

☐ Find the missing number by counting on.

11. $4 + \square = 7$

12. $5 + \square = 6$

13. $3 + \square = 5$

14. $8 + \square = 10$

15. $\square + 2 = 6$

16. $\square + 3 = 4$

17. $9 + \square = 14$

18. $7 + \square = 12$

19. $\square + 8 = 15$

☐ Find the missing number by counting on.

20. $5 + \square = 9$

21. $6 + \square = 8$

22. $6 + \square = 10$

23. $8 + \square = 10$

24. $\square + 2 = 6$

25. $\square + 3 = 4$

26. $\square + 16 = 19$

27. $7 + \square = 12$

28. $12 = 8 + \square$

29. $15 = \square + 10$

30. $\square + 14 = 19$

31. $20 = \square + 17$

32. BONUS $21 + \square = 23$

33. BONUS $42 + \square = 46$

34. BONUS $65 + \square = 68$

35. BONUS $87 + \square = 90$

☐ Count on to find the missing number.

☐ Draw shaded and unshaded circles to check your answer.

36. $3 + \square = 7$

37. $4 + \square = 8$

38. $9 = \square + 5$

OAI-58 Subtracting by Counting On

Find 5 − 2 by counting on from the smaller number.

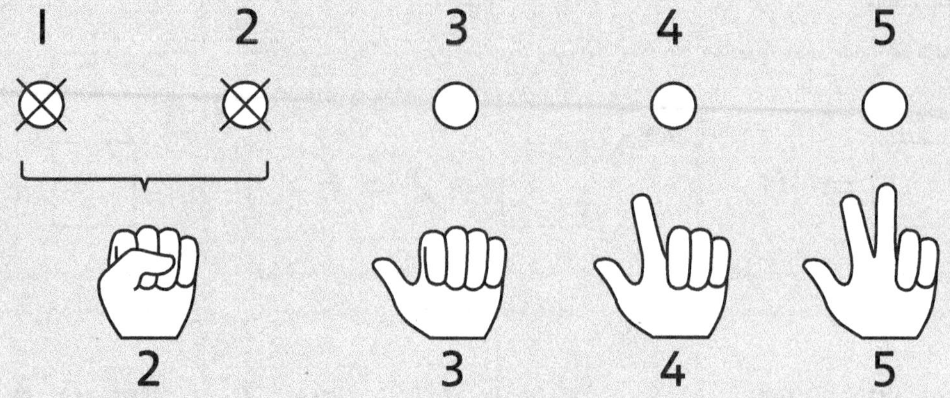

The answer is the number of fingers you have up.

$5 - 2 = 3$

☐ Subtract by counting on from the smaller number.

1. $8 - 6 = \square$
2. $9 - 5 = \square$
3. $7 - 4 = \square$
4. $9 - 8 = \square$
5. $8 - 4 = \square$
6. $6 - 3 = \square$
7. $7 - 5 = \square$
8. $10 - 6 = \square$
9. $12 - 10 = \square$
10. $15 - 11 = \square$
11. $19 - 17 = \square$
12. $19 - 12 = \square$

☐ Subtract by counting on.

13. $11 - 8 = \square$

14. $17 - 15 = \square$

15. $13 - 10 = \square$

16. $20 - 18 = \square$

17. $\square = 9 - 4$

18. $\square = 16 - 13$

19. $\square = 19 - 14$

20. $\square = 10 - 6$

21. $20 - 16 = \square$

☐ Count on to subtract.

22. $11 - 6 = \square$

23. $17 - 13 = \square$

24. $\square = 13 - 8$

25. $\square = 20 - 15$

26. $19 - 16 = \square$

27. $17 - 11 = \square$

28. $20 - 13 = \square$

29. $18 - 15 = \square$

30. $18 - 10 = \square$

31. $\square = 19 - 11$

32. $\square = 20 - 11$

33. $\square = 18 - 9$

☐ Subtract by counting on.

34. 13 − 7 = ☐

35. ☐ = 18 − 14

36. 15 − 9 = ☐

37. ☐ = 16 − 8

38. 12 − 9 = ☐

39. ☐ = 16 − 14

☐ Write the subtraction sentence to find the answer.

40. Jill has 3 pears.
She eats 2 of them.
How many does she have left? *3 − 2 = 1*

41. Ava has 20 stamps.
She gives 17 to her friends.
How many does she have left? ____________________

42. 17 frogs are on a log. 13 hop off.
How many are still on the log? ____________________

43. BONUS

You have 12 marbles. You give some to your partner.
Now you have 8 marbles.
How many did you give away? ____________________

OAI-59 Number Sentences (3)

☐ Find the missing number by adding.

1. $3 + 5 = \boxed{8}$
2. $6 + 2 = \square$
3. $8 + 3 = \square$
4. $\square = 5 + 4$
5. $\square = 9 + 4$
6. $6 + 6 = \square$

☐ Find the missing number by counting on.

7. $2 + \boxed{7} = 9$
8. $6 + \square = 8$
9. $5 = 2 + \square$
10. $9 + \square = 10$
11. $18 = \square + 14$
12. $2 + \square = 8$

☐ Find the missing number by adding or counting on.

13. $5 + 4 = \square$
14. $\square + 15 = 19$
15. $2 + 14 = \square$
16. $\square + 11 = 17$
17. $12 + \square = 14$
18. $18 + \square = 20$
19. $16 + 1 = \square$
20. $11 = \square + 7$
21. $17 + 3 = \square$

☐ Is the number sentence **true** or **false**?

22. $7 + 2 = 8$

true (false)

23. $10 + 2 = 12$

true false

24. $5 + 3 = 10$

true false

25. $8 + 3 = 11$

true false

26. $6 = 4 + 3$

true false

27. $17 = 3 + 14$

true false

28. BONUS $7 - 2 = 5$

true false

29. BONUS $9 - 4 = 6$

true false

☐ Write the answer.

30. Jayden writes $3 + 7 = 10$.

Is he correct? How do you know?

31. 8 rabbits are in a field.

6 hop away.

Write a number sentence to find how many are left.

OAI-60 Identifying Parts and Totals

☐ What things are you adding?

1.

3 big frogs

2 small frogs

frogs

2.

2 red marbles

4 green marbles

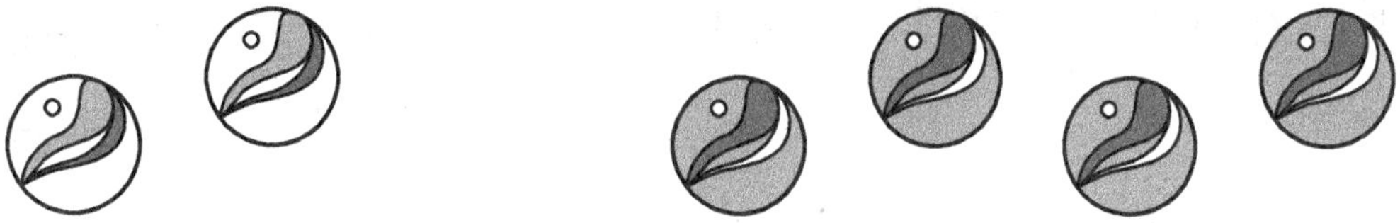

____

3.

3 new pencils

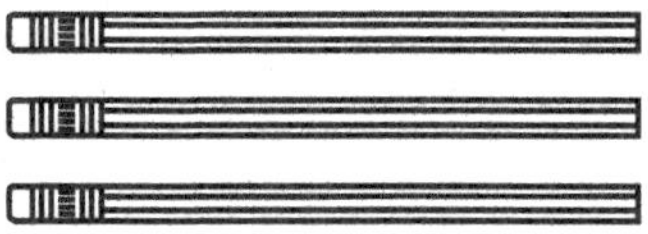

5 used pencils

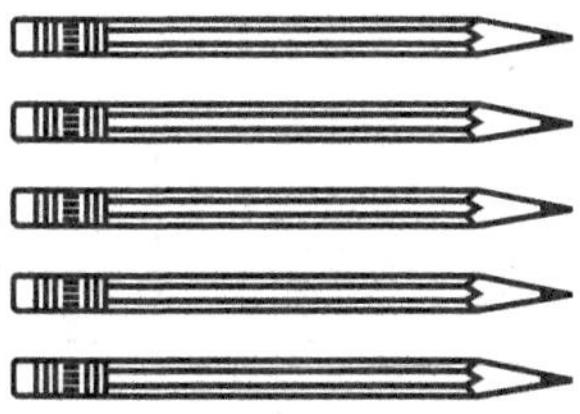

____

4.

4 big keys

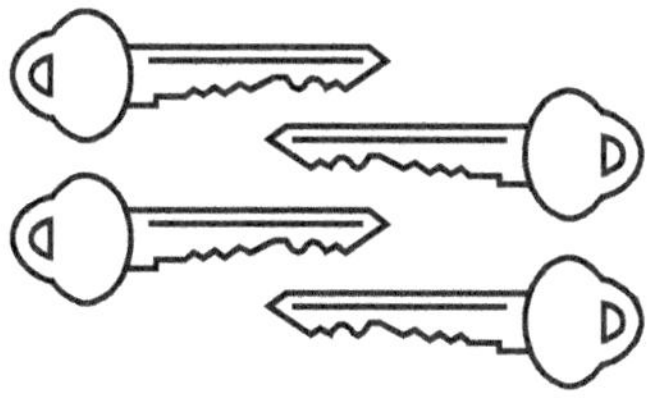

7 small keys

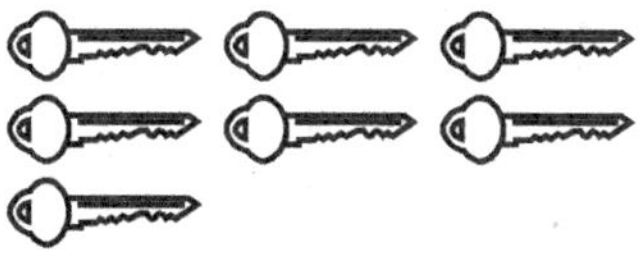

____

5.

6 green apples

6 red apples

____

☐ Add.

☐ Write what you are adding.

6. There are 5 big frogs.

There are 2 small frogs.

There are __7__ __frogs__ altogether.

7. There are 4 new pencils.

There are 2 used pencils.

There are _____ ____________ altogether.

8. There are 5 green apples.

There are 5 red apples.

There are _____ ____________ in total.

9. There are 3 empty mugs.

There are 2 full mugs.

There are _____ ____________ in total.

☐ Make a problem for the picture.
Use words from the box.

~~big~~	empty	full
small	happy	sad

10.

There are __3__ __big__ frogs.

There are __2__ ______ frogs.

There are __5__ frogs altogether.

11.

There are ____ ______ bowls.

There are ____ ______ bowls.

There are __5__ bowls in total.

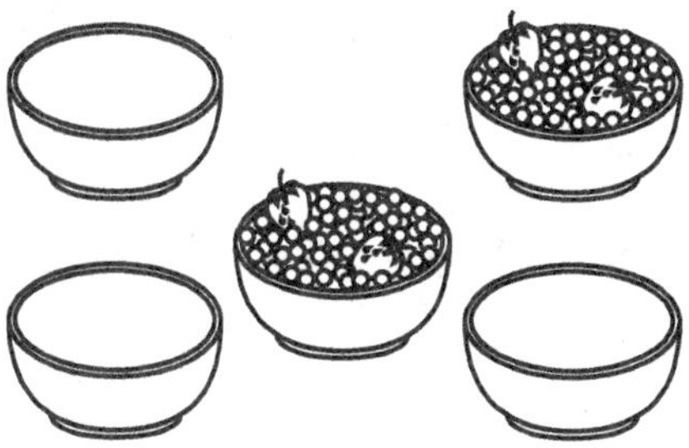

12.

There are ____ ______ faces.

There are ____ ______ faces.

There are ____ faces in total.

OAI-61 Drawing Pictures to Find Parts and Totals

There are 5 apples.

3 are red. The rest are green.

How many are green?

Draw 5 circles to show 5 apples.

Shade 3 circles to show 3 red apples.

There are 2 green apples.

There are red and green apples.

☐ Draw a picture to find how many are green.

1. There are 4 apples.
3 are red.

___I___ green apple

2. There are 5 apples.
4 are red.

______ green apple

3. There are 3 apples.
I is red.

______ green apples

4. There are 6 apples.
2 are red.

______ green apples

There are 3 red apples and 4 green apples.
How many apples are there in total?

Draw 3 red apples.

Draw 4 green apples.

There are 7 apples.

☐ Draw a picture to find the total number of apples.

5. There are 2 red apples and 3 green apples.

5 apples

6. There are 2 red apples and 2 green apples.

______ apples

7. There are 3 red apples. There are 2 green apples.

______ apples

8. Jane has 2 red apples and 1 green apple.

______ apples

9. There are 4 red apples and 2 green apples in a bowl.

______ apples

10. Don has 3 green apples and 1 red apple.

______ apples

☐ Draw a picture to find the answer.

11. Tony has 3 red apples and 3 green apples. How many apples does he have?

_____ apples

12. Kate has 5 apples. 3 are red. The rest are green. How many are green?

_____ green apples

13. 3 red apples and 1 green apple fall out of a tree. How many apples fall?

_____ apples

14. A bowl has 2 green apples and 3 red apples. How many apples are in the bowl?

_____ apples

15. A bowl has 8 apples. 3 are green. The rest are red. How many are red?

_____ red apples

16. A bowl has 5 red apples and 4 green apples. How many apples are in the bowl?

_____ apples

Draw a picture to find the answer.

17. Blanca has 3 yellow balls and 2 purple balls. How many balls does she have?

______ balls

18. Roy has 5 pets. 3 are fish. The rest are cats. How many cats does he have?

______ cats

19. BONUS Amy sees 12 birds. 7 are robins. The rest are jays. How many jays does she see?

______ jays

20. 7 children are in a pool. 3 are girls. How many are boys?

21. 3 boys and 6 girls are in a pool. How many children are in the pool?

Parts and Totals

☐ Draw circles for the apples.

1. 6 apples

○○○○○○	
○○	○○○○

2 red 4 green

2. 5 apples

○○○○○	

1 red 4 green

3. 4 apples

2 red 2 green

4. 6 apples

3 red 3 green

5. 5 apples

2 red 3 green

6. 3 apples

1 red 2 green

7. 4 apples

3 red 1 green

8. 2 apples

1 red 1 green

Draw circles to find the missing number.

9.

5 circles

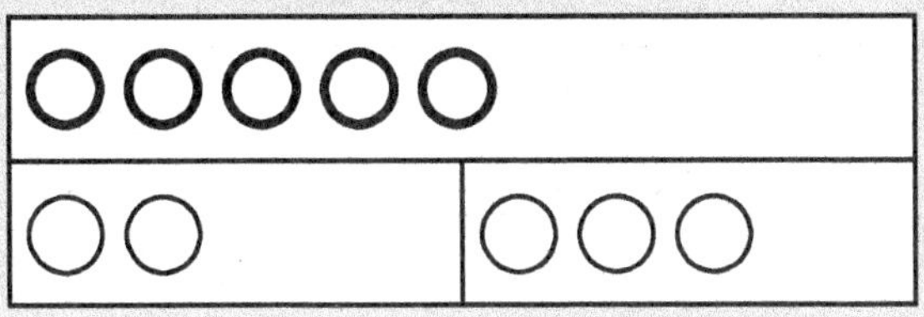

2 red 3 green

10.

5 circles

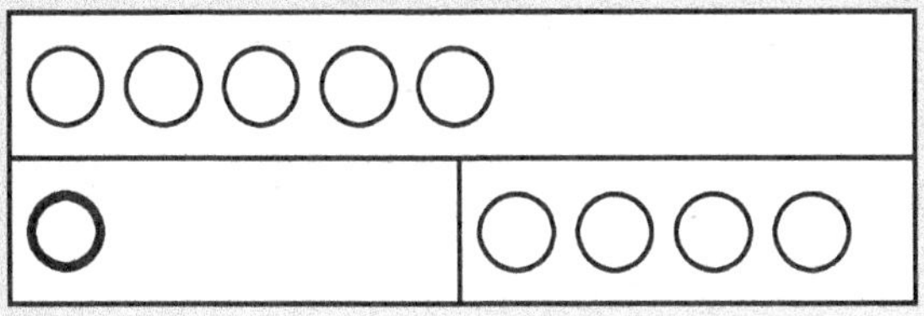

1 red 4 green

11.

_____ circles

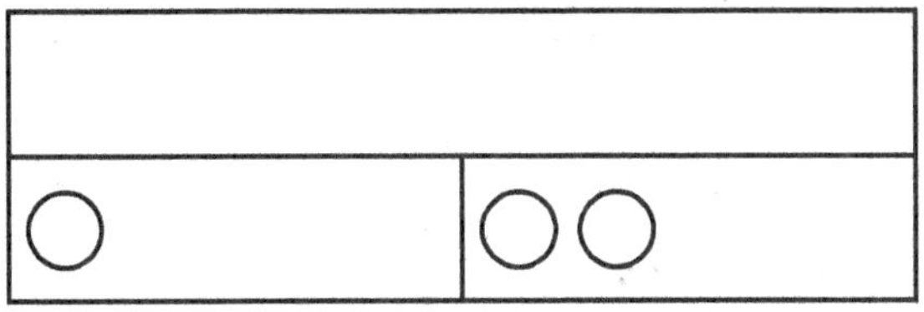

1 red 2 green

12.

_____ circles

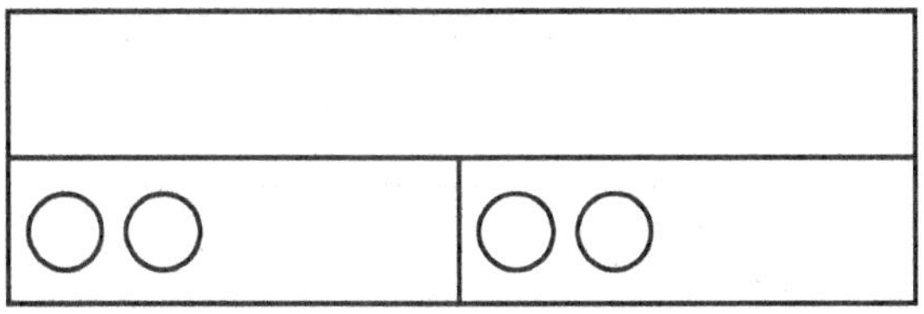

2 red 2 green

13.

4 circles

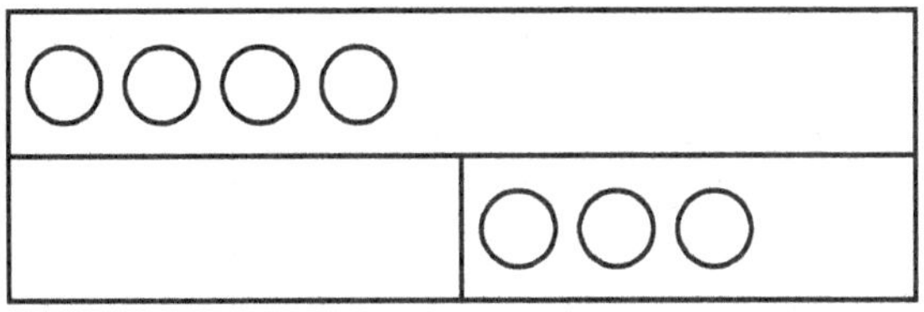

_____ red 3 green

14.

6 circles

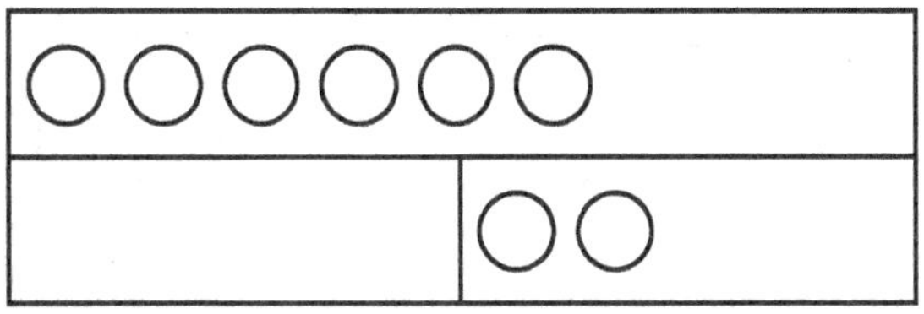

_____ red 2 green

15.

2 circles

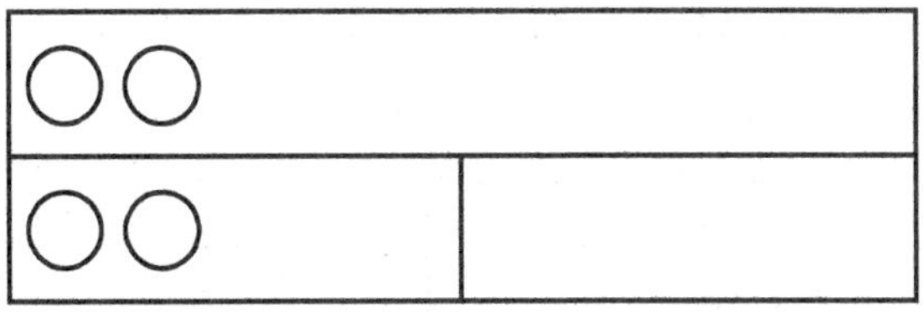

2 red _____ green

16.

5 circles

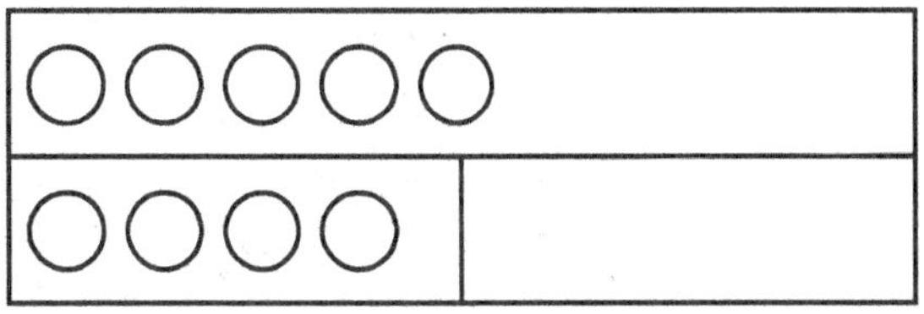

4 red _____ green

☐ Find the missing number.

17.

total	
7	
2	5
part	part

18.

total	
5	
2	3
part	part

19.

total	
3	1
part	part

20.

total	
6	
2	
part	part

21.

total	
7	1
part	part

22.

total	
7	
	3
part	part

23.

total	
4	5
part	part

24.

total	
10	
	2
part	part

☐ Find the missing number.

25.

total	
6	
3	
part	part

26.

total	
5	
	1
part	part

27.

total	
5	3
part	part

28.

total	
7	
	5
part	part

29.

total	
2	10
part	part

30.

total	
8	
7	
part	part

31. BONUS

total	
13	6
part	part

32. BONUS

total	
20	
	14
part	part

OAI-63 Parts, Totals, and Number Sentences

☐ Write an addition sentence for the picture.

1.

6	
4	2

$4 + 2 = 6$

2.

7	
3	4

____

3.

9	
4	5

____

4.

13	
10	3

____

☐ Write an addition sentence for the picture.

☐ Draw ☐ for the missing number.

5.

5	
3	

$3 + \square = 5$

6.

7	
4	

____

7.

8	
5	

____

8.

9	
	2

____

☐ Write an addition sentence for the picture.

9.

8	
	5

☐ + 5 = 8

10.

4	
	2

____

11.

7	
	3

____

12.

12	
	3

____

13.

6	3

6 + 3 = ☐

14.

4	5

____

15.

1	9

____

16.

14	3

____

☐ Fill in the missing numbers.

17.

9	
2	7

2 + [7] = 9

18.

8	
3	5

3 + 5 = [8]

19.

6	
	2

[ ] + 2 = 6

20.

5	
3	

3 + [ ] = 5

21.

2	8

2 + 8 = [ ]

22.

9	
4	

4 + [ ] = 9

☐ Write a number sentence.

☐ Fill in the missing numbers.

23. BONUS

8	
	2

____

24. BONUS

3	4

____

OAI-64 Word Problems with Parts and Totals

☐ Write **?** for the missing number.

☐ Fill in the picture.

1.

Ken has 6 blue fish and 5 green fish.

How many fish does he have?

total

?	
6	5
part	part

2.

Lynn has 3 red pens and 2 blue pens.

How many pens does she have in all?

total

part	part

3.

There are 6 boys and 7 girls in a class.

How many students are in the class?

total

part	part

☐ Write **?** for the missing number.
☐ Fill in the picture.

4.

Ivan has 5 crayons.
2 of them are red.
The rest are yellow.

How many are yellow?

total

5	
2	?

part part

5.

7 birds are on a branch.
3 of them are blue.
The rest are red.

How many are red?

total

part part

6.

14 students are on a bus.
9 of them are girls.

How many are boys?

total

part part

☐ Fill in the picture.

☐ Find the answer.

7. Hanna has 3 red apples and 4 green apples.
How many apples does she have?

___7___ apples

total	
7	
3	4
part	part

8. Ravi has 8 animal stickers.
5 are lions. The rest are tigers.
How many tiger stickers does he have?

_____ tiger stickers

total	
part	part

9. Grace finds pink shells and white shells.
She finds 8 shells. 6 are white.
How many are pink?

_____ pink shells

total	
part	part

10. Jon has 12 fish and 2 dogs.
How many pets does he have?

_____ pets

total	
part	part

☐ Fill in the picture.

☐ Find the answer.

11.

Zara puts 5 dimes in her bank.
Then she puts in 6 dimes.
How many dimes does she put in her bank altogether?

_____ dimes

total

part part

12.

17 children play tag.
7 are boys.
How many are girls?

_____ girls

total

part part

☐ Fill in the picture.

☐ Write the number sentence.

☐ Find the answer.

13. BONUS

7 girls and 9 boys play baseball.
How many children play baseball?

____

_____ children play baseball

total

part part

OAI-65 More Number Sentences (Advanced)

☐ Circle the number sentence you would use to solve the problem.

1. 5 cats are in a basket. 3 more get in.

How many cats are in the basket now?

5 + 3 = ☐   3 + ☐ = 5

2. Abdul buys 3 turtles. Then he buys some lizards.

He buys 5 animals in total.

How many lizards does he buy?

5 + 3 = ☐   3 + ☐ = 5

☐ Write a number sentence.

☐ Fill in the answer.

3. Jay has 3 pens. Then he buys some more pens.

Now he has 7 pens. How many does he buy?

3 + [4] = 7

4. 3 frogs are in a pond. 5 more jump in.

How many frogs are in the pond?

____

☐ Write a number sentence.

☐ Find the answer.

5.

Ted picks 6 flowers in a field.

He picks some more in a forest.

He picks 10 flowers in all.

How many does he pick in the forest? ________________

6.

Marta sees 7 ducks on a pond.

Then 2 more ducks land on the pond.

How many ducks does she see now? ________________

7.

Helen eats 5 green grapes.

Then she eats some red grapes.

She eats 9 grapes altogether.

How many red grapes does she eat? ________________

8.

6 girls are in a class.

There are 10 students in the class.

How many boys are in the class? ________________

OAI-66 Subtracting and How Many More

☐ Draw circles to make 5.

☐ Find the missing number.

1. 

5 is __1__ more than 4.

2. 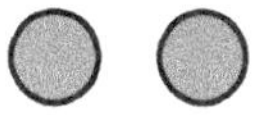

5 is _____ more than 2.

3. 

5 is _____ more than 1.

4. 

5 is _____ more than 3.

☐ Draw circles to find the missing number.

5. 

6 is __4__ more than 2.

6. 

6 is _____ more than 3.

7. 

7 is _____ more than 3.

8. 

4 is _____ more than 1.

9. 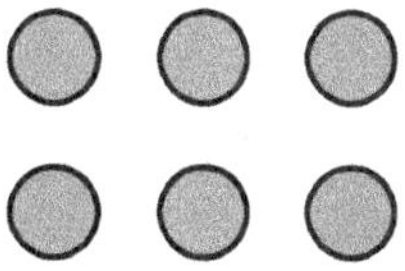

10 is _____ more than 6.

10. 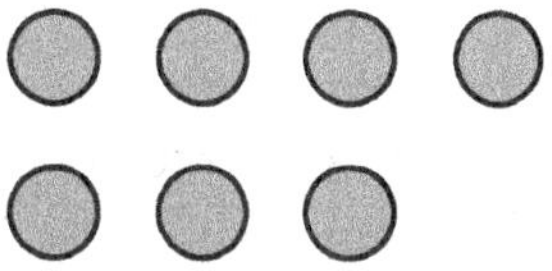

12 is _____ more than 7.

☐ Shade how many to take away. Then subtract.

11.

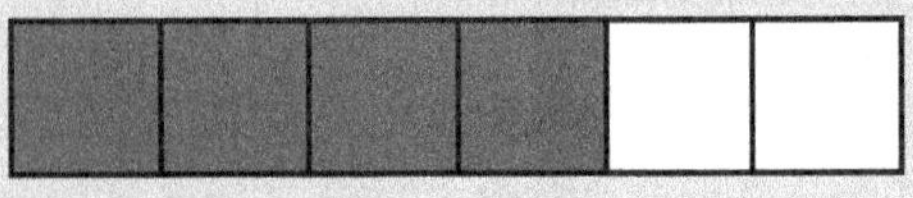

6 − 4 = 2

12.

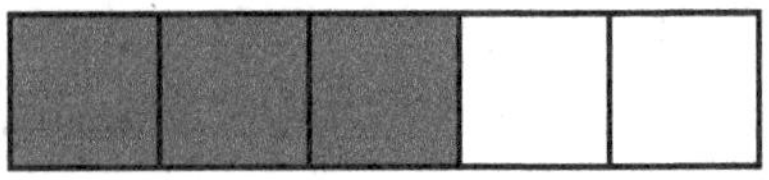

5 − 3 = ____

13.

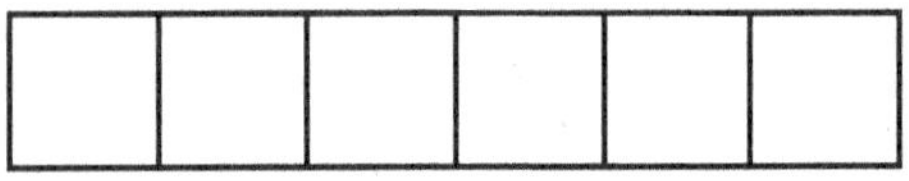

6 − 2 = ____

14.

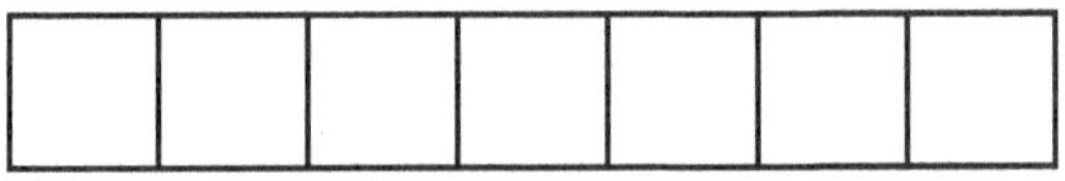

7 − 2 = ____

☐ Shade the squares that match.

☐ Count how many more.

☐ Fill in the blank.

15.

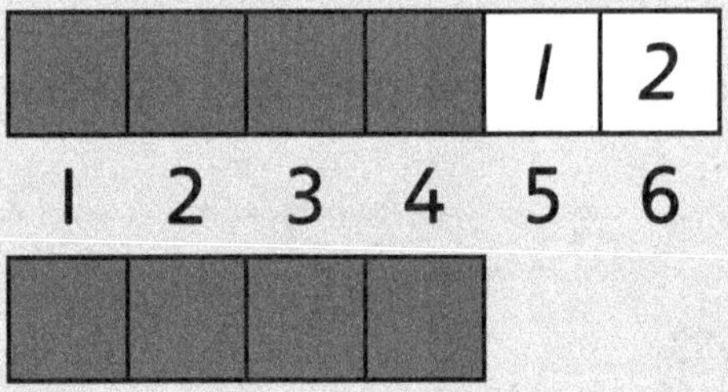

6 is 2 more than 4.

16.

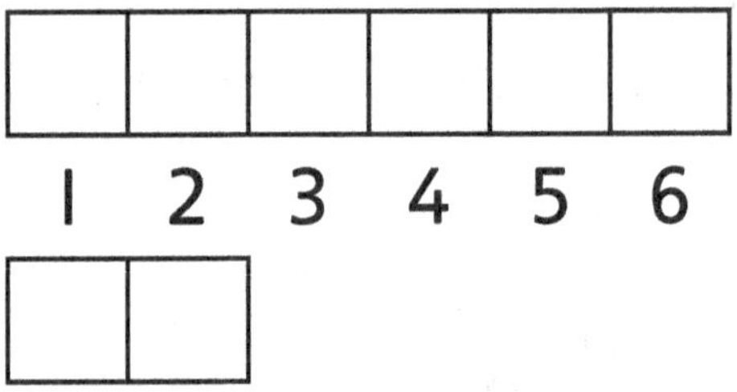

6 is ____ more than 2.

17.

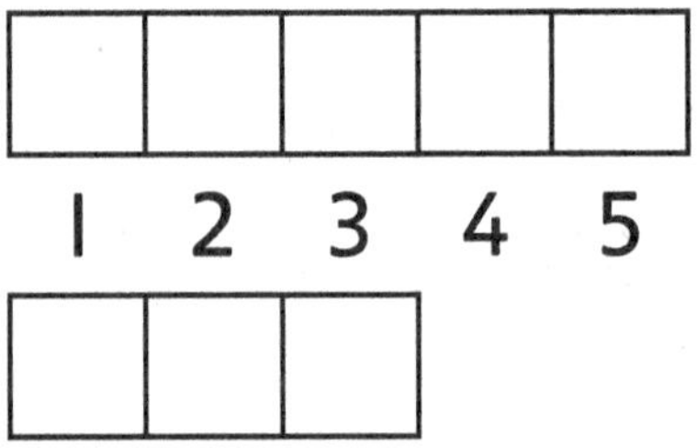

5 is ____ more than 3.

18.

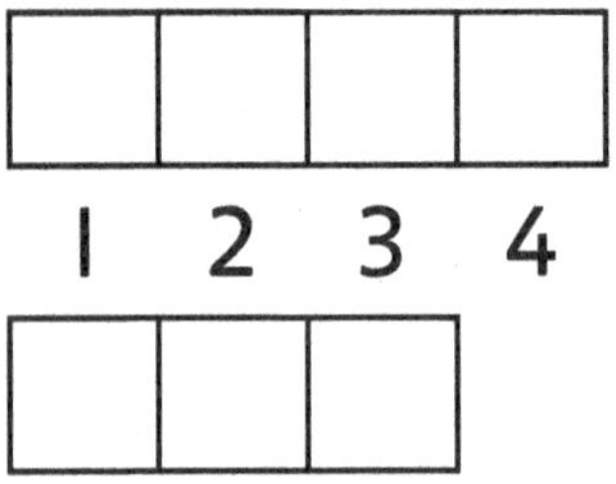

4 is ____ more than 3.

☐ Shade the number of squares that match.

☐ Count how many more. Then subtract.

19.

1 2 3 4 5 6 7

7 is __3__ more than 4.

7 − 4 = __3__

20.

1 2 3 4 5 6

6 is _____ more than 1.

6 − 1 = _____

21.

1 2 3 4 5 6 7

7 is _____ more than 3.

7 − 3 = _____

22.

1 2 3 4 5

5 is _____ more than 2.

5 − 2 = _____

23.

1 2 3 4 5 6 7 8 9

9 is _____ more than 4.

9 − 4 = _____

OAI-67 More Than with Pictures

- [ ] Draw circles to show how many.
- [ ] Fill in the answer.

1.

Mike has 4 stickers.

Sharon has 2 more stickers than Mike.

Mike	O	O	O	O			
Sharon	O	O	O	O	O	O	

Sharon has __6__ stickers.

2.

Ray has 3 toy boats.

Beth has 4 more toy boats than Ray.

Ray							
Beth							

Beth has _____ toy boats.

3.

Tess eats 2 strawberries.

Kyle eats 4 more strawberries than Tess.

Tess							
Kyle							

Kyle eats _____ strawberries.

☐ Draw circles to show how many.

☐ Fill in the answer.

4.

Sun sees 3 birds.

Glen sees 3 more birds than Sun.

Sun								
Glen								

Glen sees _____ birds.

5.

Josh has 2 hats.

Marla has 5 more hats than Josh.

Josh								
Marla								

Marla has _____ hats.

6.

Randi has 6 books.

Bev has 2 more books than Randi.

Randi								
Bev								

Bev has _____ books.

☐ Draw O to show how many.

☐ Draw X to show how many fewer.

7. Ethan has 6 stickers. Emma has 2 fewer stickers.

Ethan	O	O	O	O	O	O	
Emma	O	O	O	O	X	X	

Emma has __4__ stickers.

8. Karen paints 7 pictures. Fred paints 3 fewer pictures.

Karen							
Fred							

Fred paints _____ pictures.

☐ Who has more blocks?

☐ Draw O and X to show how many blocks each person has.

9. Ben has 3 blocks. He has 2 fewer blocks than Clara.

10. Dan has 5 blocks. He has 3 more blocks than Zara.

11. Anna has 10 blocks. She has 4 more blocks than Kim.

OAI-68 Comparing Using Subtraction

☐ Use the word sentence to subtract.

1. 6 is 2 more than 4.

   6 − 4 = __2__

2. 6 is 4 more than 2.

   6 − 2 = ____

3. 12 is 7 more than 5.

   12 − 5 = ____

4. 25 is 2 more than 23.

   25 − 23 = ____

☐ Subtract.

5. 7 is how many more than 5?

   __7__ − __5__ = ____

6. 10 is how many more than 4?

   __10__ − ____ = ____

7. 9 is how many more than 6?

   __9__ − ____ = ____

☐ Write the subtraction.

8. Amy has 6 pears. Bo has 4 pears.

Amy has __6 − 4__ more pears.

9. Marco has 5 pencils. Lily has 4 pencils.

Marco has __5 −__ more pencils.

10. Jen has 10 marbles. Ed has 8 marbles.

Jen has ________ more marbles.

11. Mindy has 5 books. Alan has 2 books.

Mindy has ________ more books.

12. 6 peaches are on a plate. 3 peaches are in a bowl.

The plate has ________ more peaches.

13. BONUS Will is 6 years old. Grace is 8.

Grace is ________ years older.

☐ Write the subtraction sentence.

14. Vicky has 9 stamps. Jon has 5 stamps.

How many more stamps does Vicky have?

$9 - 5 = 4$

15. Tasha has 8 toy cars. Mark has 5 toy cars.

How many more cars does Tasha have?

16. Alex has 9 beads. Tessa has 7 beads.

How many more beads does Alex have?

17. Tom has 17 balloons. Hanna has 7 balloons.

How many more balloons does Tom have?

18. Kim has 11 markers. Ron has 6 markers.

How many more markers does Kim have?

19. BONUS Nancy takes 17 photos. Cam takes 20 photos.

How many more photos does Cam take?

OAI-69 Choosing to Add or Subtract

☐ Show how you find the answer.

1. Tim has 3 pears.
He eats 2 of them.
How many pears does he have left?

3 (−) 2

2. There are 3 sharks in a tank.
There are 4 sharks in another tank.
How many sharks are there altogether?

____ ◯ ____

3. Kathy eats 5 oranges.
Sam eats 2 oranges.
How many more oranges does Kathy eat?

____ ◯ ____

4. Tina buys 3 fish.
Then she buys 5 more.
How many fish does she buy altogether?

____ ◯ ____

☐ Write a number sentence to solve the problem.

5. Sal has 5 red pencils
and 3 blue pencils.
How many pencils does he have?

____ ◯ ____ = ____

6. Mona brings 5 footballs to practice.
Rick brings 3 footballs.
How many footballs do they bring altogether?

____ ◯ ____ = ____

7. 8 birds are on a wire.
4 fly away.
How many birds are left on the wire?

____ ◯ ____ = ____

8. Eddy reads 18 books in a month.
Alice reads 15 books.
How many more books does Eddy read?

9. BONUS

Jane collects 24 stamps. Sam collects 32 stamps.
How many stamps do they collect altogether?

☐ Fill in the number sentence.

10. Milly buys 3 carrots. Don buys 4 tomatoes.
Cathy buys 5 beets.

How many vegetables do they buy?

3 (+) 4 (+) 5 = 12

11. Carlos has 5 bunnies. Sara has 3 bunnies.
Peter has 2 bunnies.

How many bunnies do they have altogether?

____ ◯ ____ ◯ ____ = ____

12. Anna's class has 2 mice, 3 fish, and 3 turtles.
How many animals does her class have?

13. Ron has 10 apples. He eats 2 apples.
Then he gives 2 apples to his friends.
How many apples does Ron have now?

14. BONUS

Tony has 5 leaves. He finds 2 more.
Then he gives 3 to his friend.
How many leaves does Tony have now?

Solving Number Sentences (Advanced)

What number can you add to 8 to make 14?

Add to make 10.

Add more to make 14.

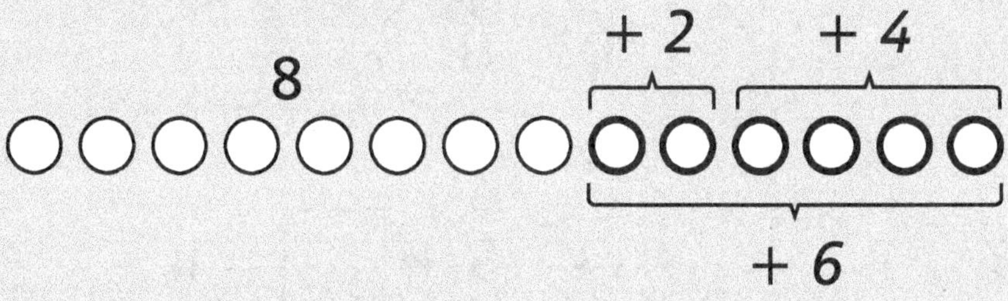

8 + [6] = 14

☐ Fill in the missing numbers.

1. 7 →(+3)→ 10 →(+4)→ 14

2. 6 →(+ ___)→ 10 →(+ ___)→ 13

3. 9 →(+ ___)→ 10 →(+ ___)→ 12

4. 5 →(+ ___)→ 10 →(+ ___)→ 17

5. 8 →(+ ___)→ 10 →(+ ___)→ 16

6. 7 →(+ ___)→ 10 →(+ ___)→ 15

☐ Fill in the missing numbers.

7.

8 —(+2)→ 10 —(+5)→ 15

8 + [7] = 15

8.

7 —(+ ___)→ 10 —(+ ___)→ 15

7 + [ ] = 15

9.

6 —(+ ___)→ 10 —(+ ___)→ 13

6 + [ ] = 13

10.

5 —(+ ___)→ 10 —(+ ___)→ 11

5 + [ ] = 11

11.

4 —(+ ___)→ 10 —(+ ___)→ 13

4 + [ ] = 13

12.

7 —(+ ___)→ 10 —(+ ___)→ 12

7 + [ ] = 12

☐ Write a related addition sentence.

☐ Fill in the missing numbers.

13.

14 − [8] = 6

6 + [8] = 14

14. BONUS

12 − [ ] = 5

___ + [ ] = ___

OAI-71 Coins

☐ Write the value of the coin.

1.

25 cents

2.

____

3.

____

4.

____

5. How much more is a dime worth than a nickel? _____¢

6. How much more is a quarter worth than a nickel? _____¢

7. How much more is a quarter worth than a penny? _____¢

☐ Circle the coin that is worth more.

8.

9.

10.

11.

12.

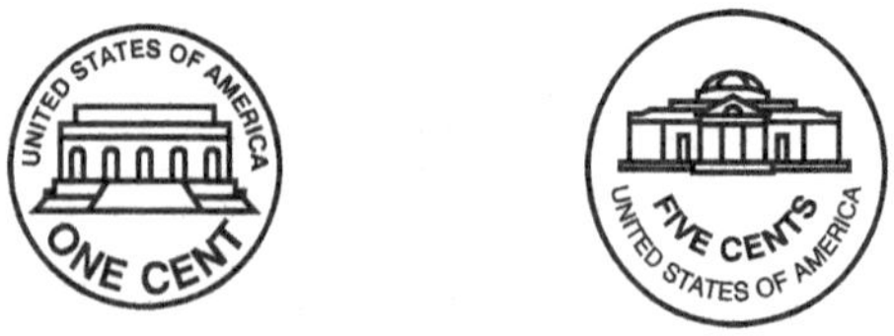

13.

☐ Circle the coin that is worth the most.

14.

15.

☐ How much money?

16.

4¢

17.

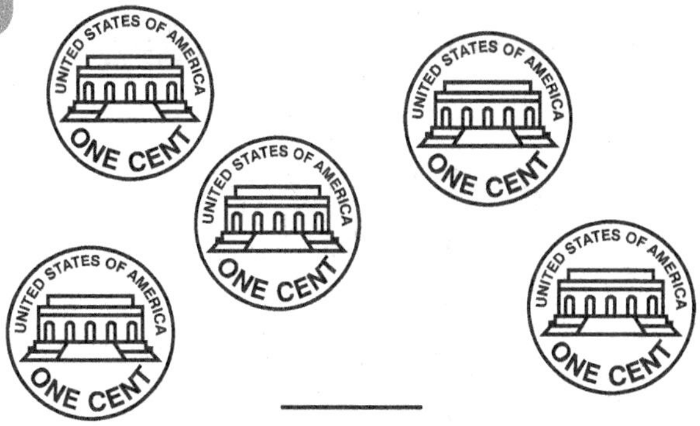

______

18.

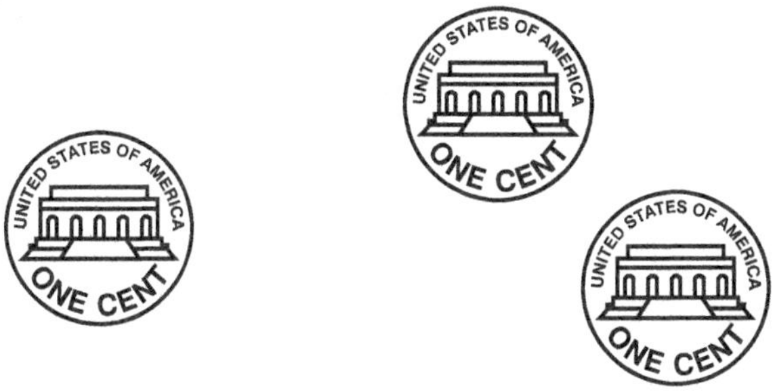

______

19.

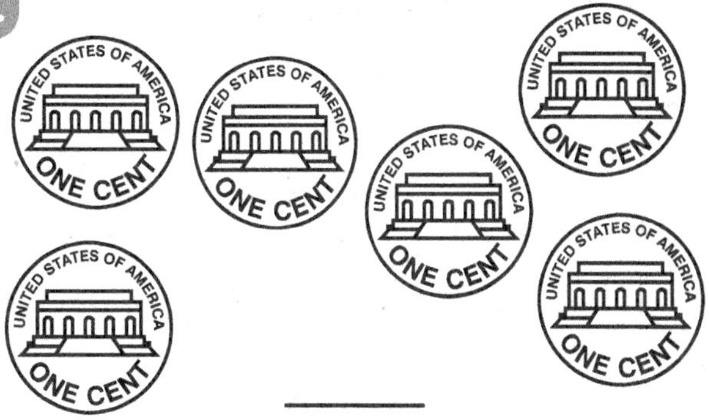

______

20.

______

21.

______

OAI-72 How Much Money?

Sandy trades 5 pennies for a nickel.

☐ Draw Sandy's money.

1.

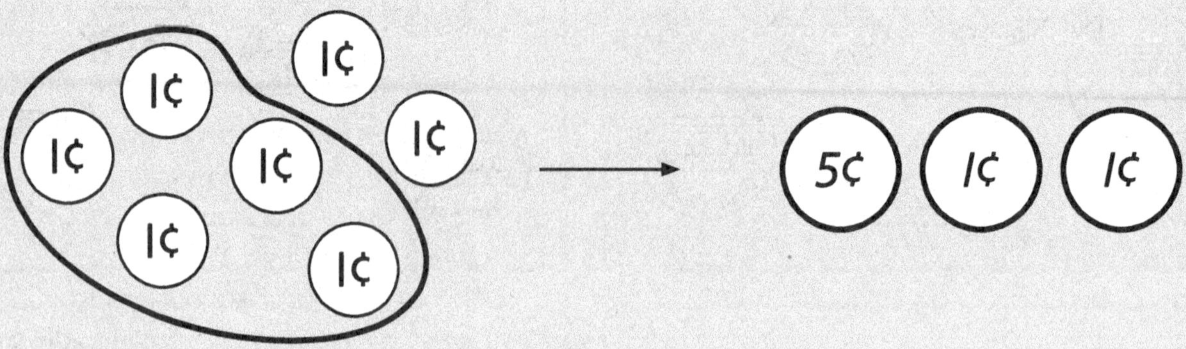

2.

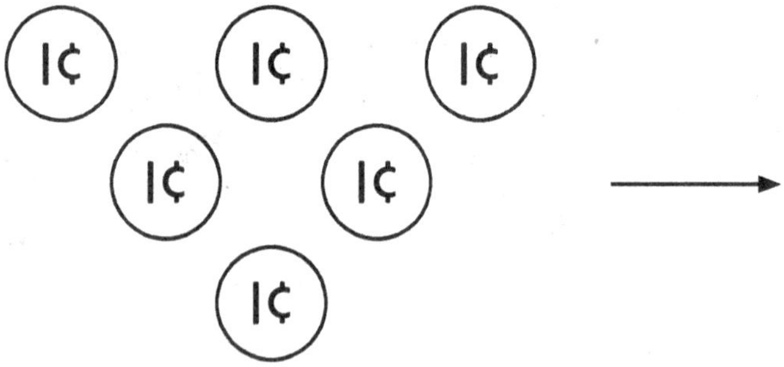

3.

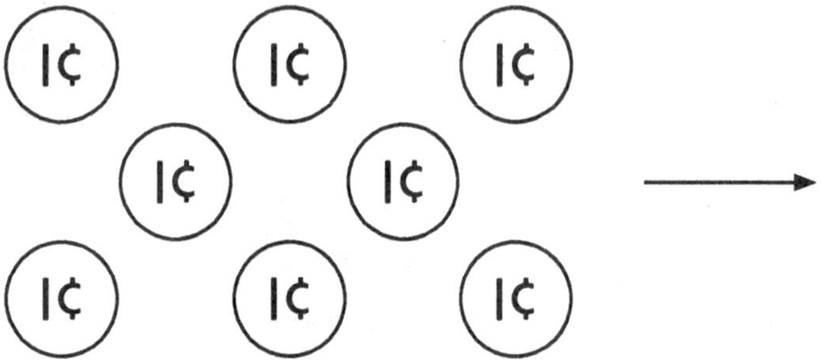

4.

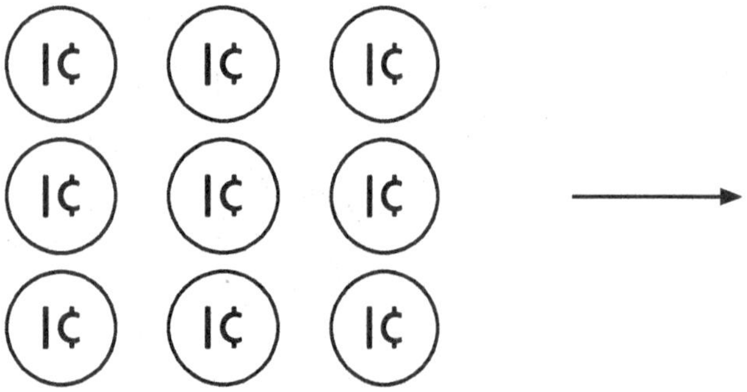

☐ Add.

5. 5 + 5 = ____

6. 10 + 5 = ____

7. 15 + 5 = ____

☐ Write an addition sentence.

☐ Add the money.

8. (5¢) (5¢)   *5 + 5* = *10¢*

9. (5¢) (5¢) (5¢)   ________ = ____

10. (5¢) (5¢) (5¢) (5¢)   ________ = ____

☐ Add the money.

11. (5¢) (1¢)   ____

12. (5¢) (1¢) (1¢) (1¢)   ____

13. (5¢) (5¢) (1¢)   ____

14. (5¢) (1¢) (1¢) (5¢)   ____

☐ Count by 1s to add pennies.

15.

1¢	1¢	1¢	1¢	1¢
1 ¢	2 ¢	____¢	____¢	____¢

☐ Add nickels.

16.

5¢	5¢	5¢	5¢	5¢
5 ¢	10 ¢	____¢	____¢	____¢

☐ Count the money.

17.

5¢	5¢	1¢	1¢	1¢
____¢	____¢	____¢	____¢	____¢

18. BONUS

5¢	5¢	5¢	1¢	1¢
____	____	____	____	____

Draw the coins you need to pay.

19.

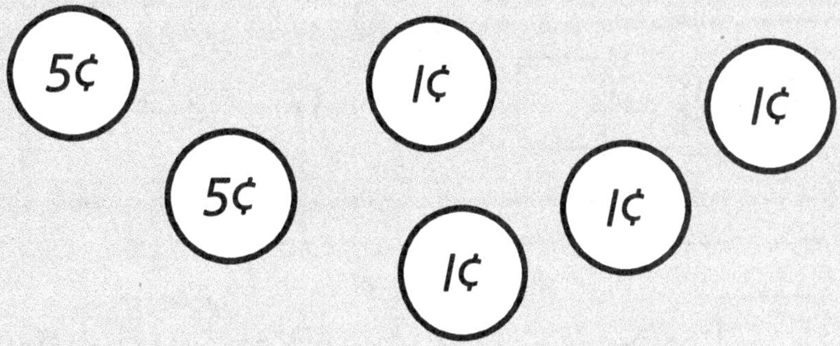

20.

21.

22.

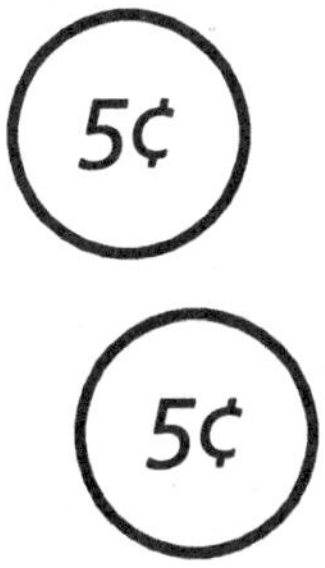

OAI-73 Review

☐ Fill in the missing number.

1. $3 + 5 = \square$

2. $17 - 2 = \square$

3. $\square + 3 = 9$

4. $\square = 8 + 4$

5. $12 = \square + 6$

6. $9 - 2 = \square$

7. $57 - 4 = \square$

8. $27 + \square = 32$

9. $68 + \square = 70$

10. BONUS $\square = 37 - 35$

11. BONUS $48 - 47 = \square$

12. BONUS $99 = 90 + \square$

☐ Find the answer.

13. It rains 8 days in May.
It rains 10 days in June.
How many more days does it rain in June?

14. Ray has 5 fish, 4 turtles, and 6 mice.
How many pets does he have?

15. Pam has 4 sisters. Jim has 3 fewer sisters.
How many sisters does Jim have?

☐ Find the answer.

16. Tess sees 12 ladybugs on the way to school.
She sees 11 more on the way home.
How many ladybugs does she see?

17. Jay picks 8 berries. Then he picks 6 more berries.
He gives 7 berries to his sister.
How many berries does he have now?

18. Some cows were in a barn.
5 more came into the barn.
Now there are 10 cows in the barn.
How many cows were in the barn?

19. Alexa says the missing number is 9.

$2 + \square = 7$

Is she correct? Explain.

20. Carlos says the missing number is 12.

$10 - 2 = \square$

Is he correct? Explain.

MDI-13 Clock Faces

This is a **clock face.**

The numbers start at 1 and end at 12.

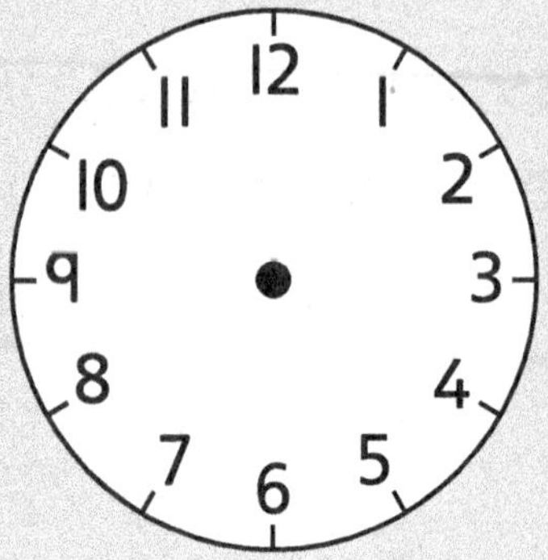

☐ Fill in the missing 3, 6, 9, or 12.

1.

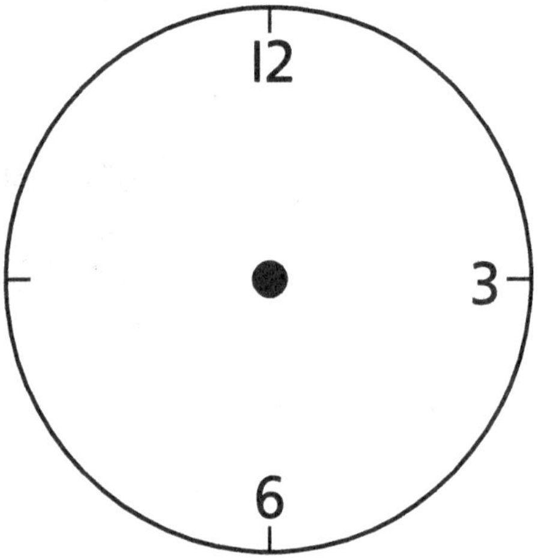

2.

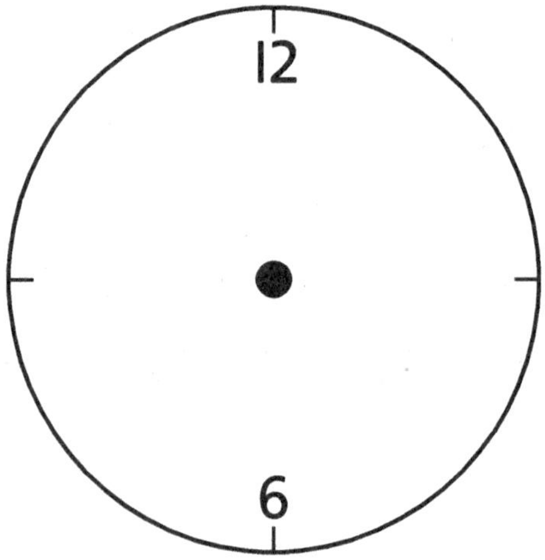

3.

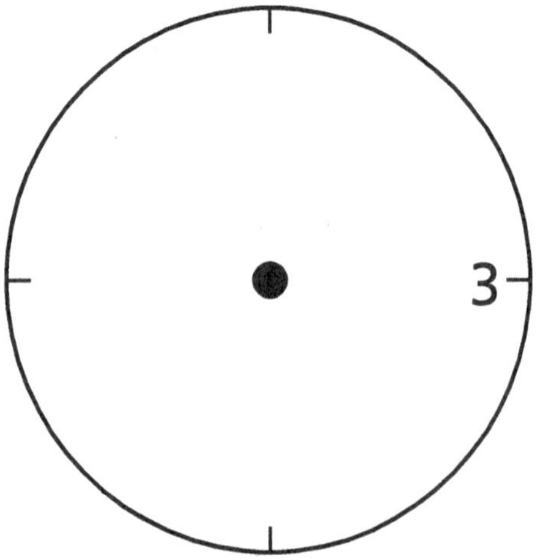

4.

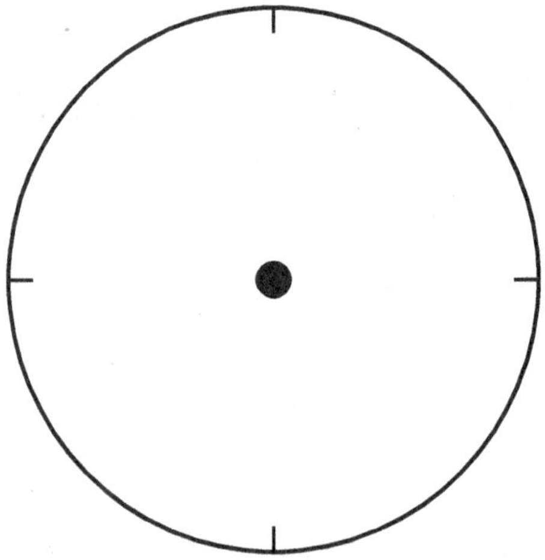

☐ Fill in the missing numbers.

5.

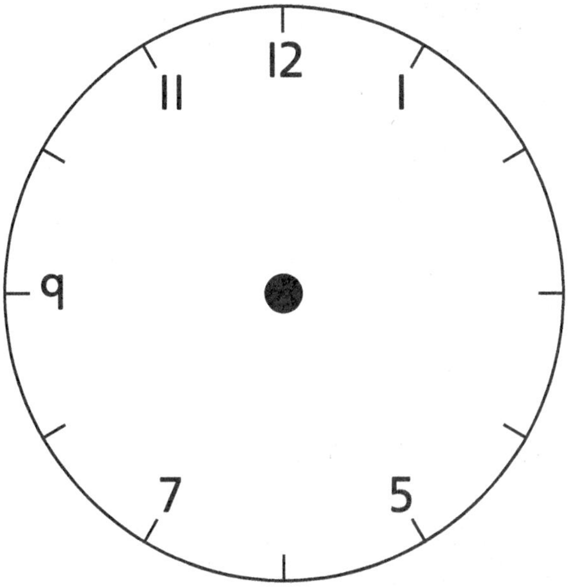

6.

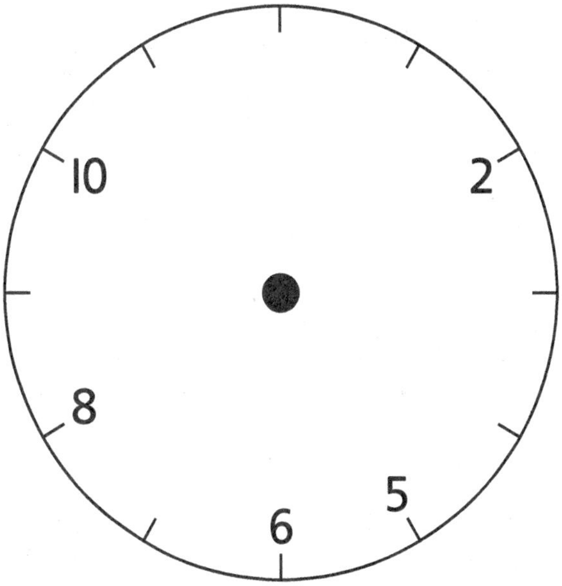

7.

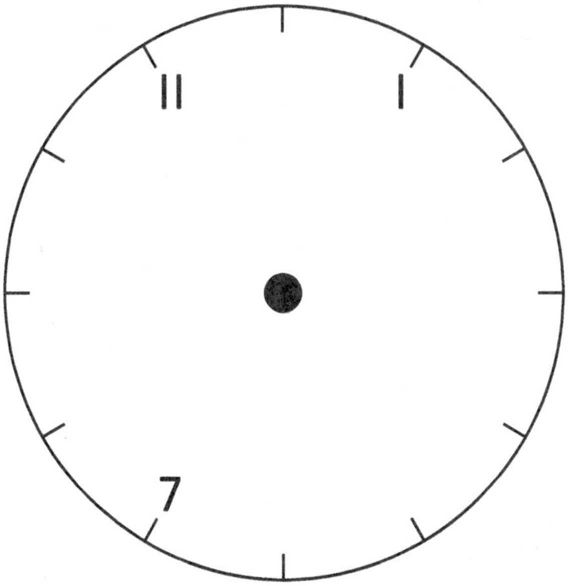

8.

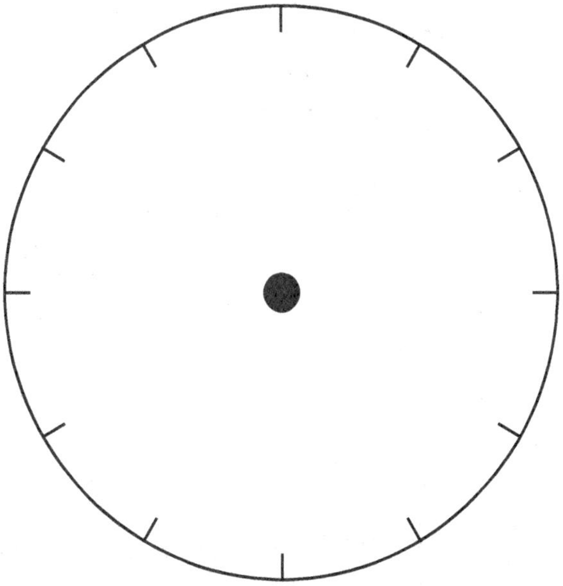

MDI-14 The Hour Hand

The **minute hand** is longer.

The **hour hand** is shorter.

☐ Circle the hour hand.

1.

2.

3.

4.

5.

6.

7. BONUS

The hour hand is pointing at the 3.
What number will it point at 1 hour later?

The hour hand

The hour hand is pointing at the 3.

The hour hand is pointing **between** 7 and 8.

☐ Where is the hour hand pointing?

8.

between

8 and 9

9.

between

____ and ____

10.

between

____ and ____

11.

between

____ and ____

12.

between

____ and ____

13.

between

____ and ____

Where is the hour hand pointing?

14.

at the

2

15.

between the

_____ and _____

16.

at the

_____

17.

at the

_____

18.

between the

_____ and _____

19.

at the

_____

MDI-15 Time to the Hour

It is **9 o'clock** or **9:00**.

☐ Write the **time**.

1.

_______ o'clock

2.

_______ o'clock

3.

_______ o'clock

4.

4 o'clock

5.

____

6.

____

7.

____

8.

____

9.

____

☐ Write the time two ways.

10.

6 o'clock

6 : 00

11.

______ o'clock

______ : 00

12.

______ o'clock

______ : 00

13.

____

______ : ______

14.

____

______ : ______

15.

____

______ : ______

16. BONUS

Jill's soccer game starts at 4:00.
The game ends 1 hour later.
What time does the game end?

☐ Draw the hands on the clock.

17. 3:00

18. 2 o'clock

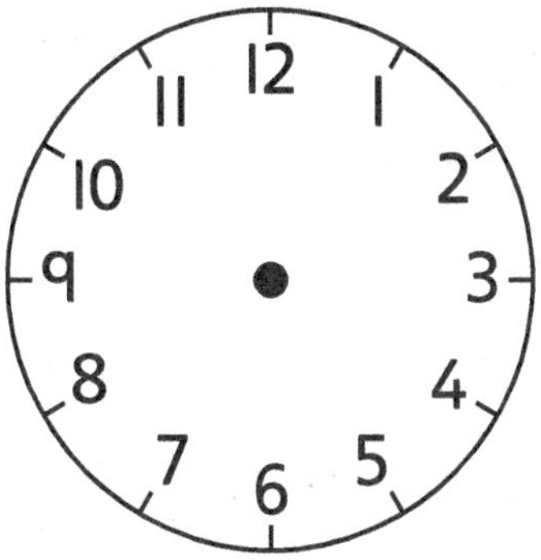

19. 5:00

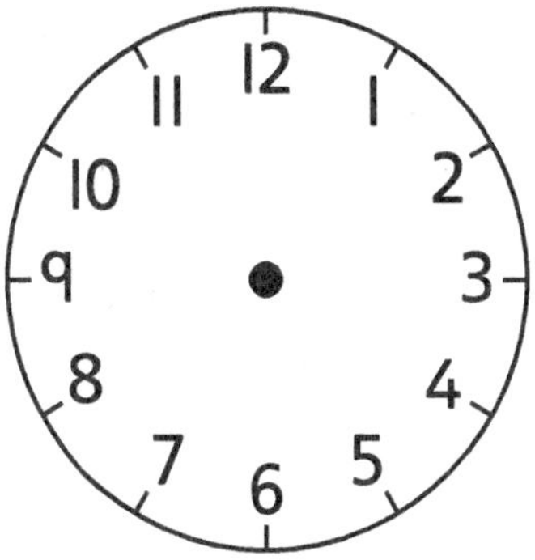

20. 11 o'clock

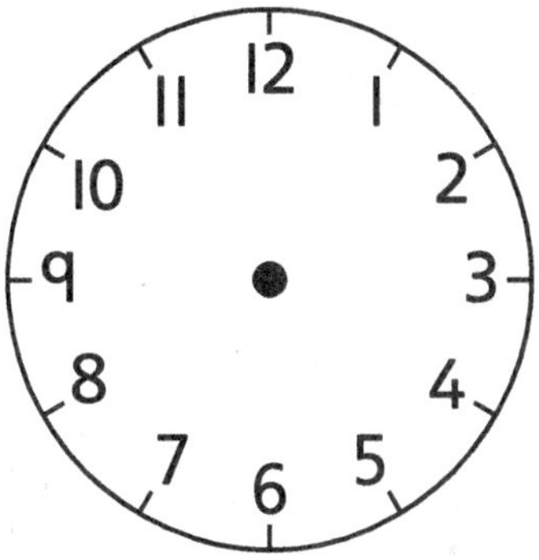

21. 7:00

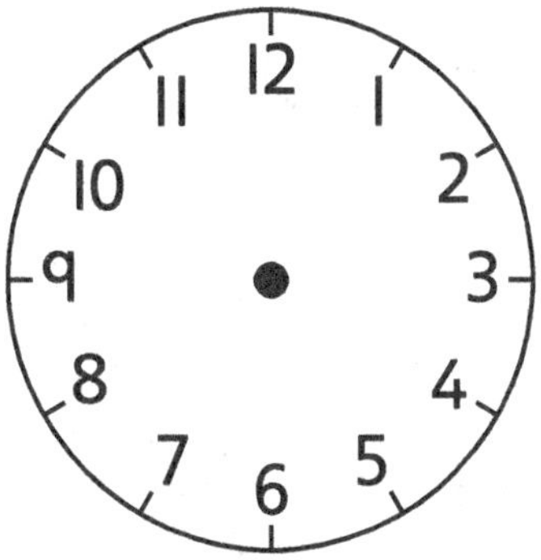

22. 12 o'clock

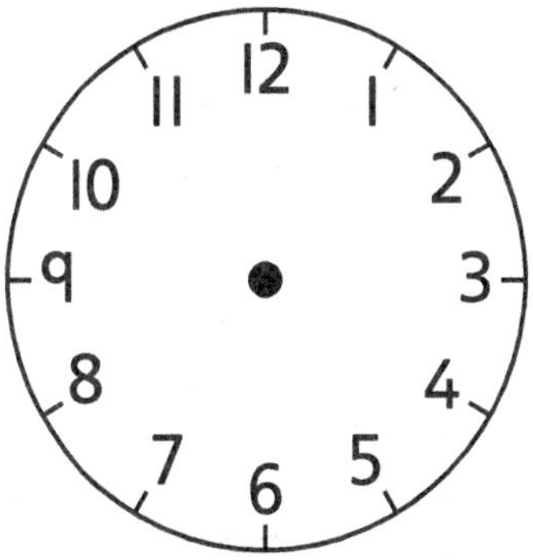

23. 7 o'clock

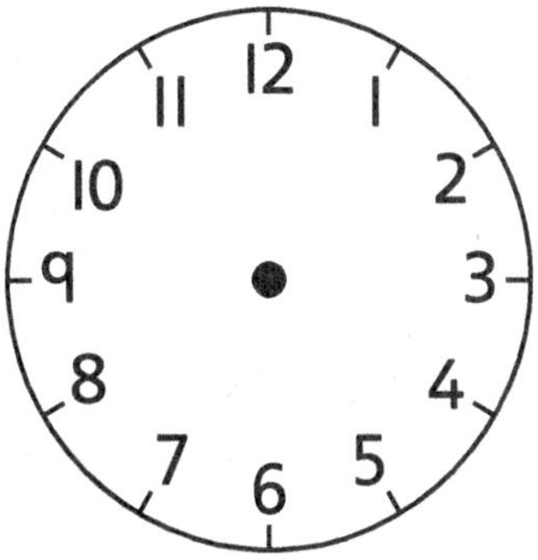

24. 6 o'clock

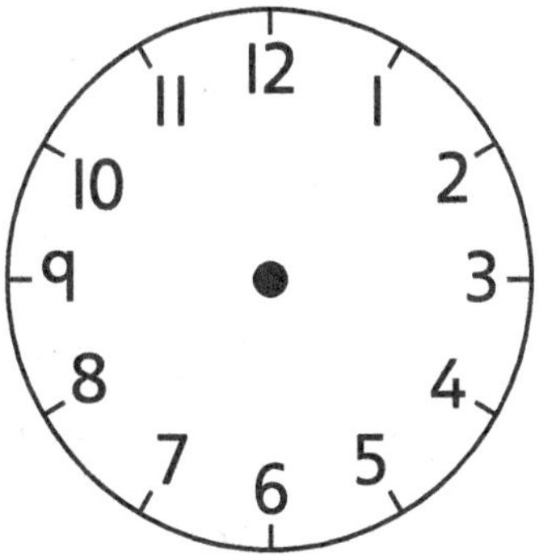

25. 9:00

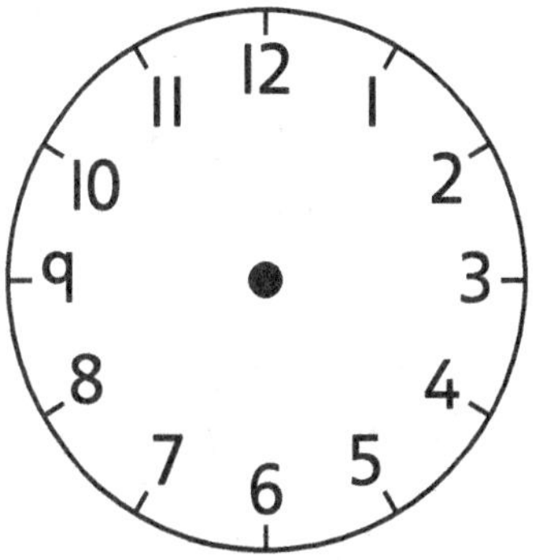

☐ **BONUS:** Circle the clocks that show the same time.

MDI-16 Time to the Half Hour

The **minute hand** moved halfway around.
The hour hand is halfway between 3 and 4.
It is **half past 3** or **3:30**.

☐ Write the time.

1.

half past ____

2.

half past ____

3.

half past ____

4.

____

5.

____

6.

____

7.

____

8.

____

9.

____

☐ Write the time two ways.

10.

half past 3

3 : 30

11.

half past ____

______ : 30

12.

half past ____

______ : 30

13.

____

______ : ______

14.

____

______ : ______

15.

____

______ : ______

16.

____

____

17.

____

____

18.

____

____

☐ Draw the hands.

19. 12:30

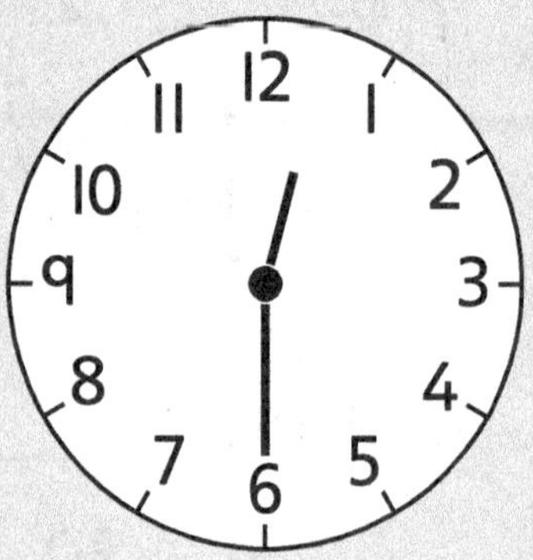

20. 2:30

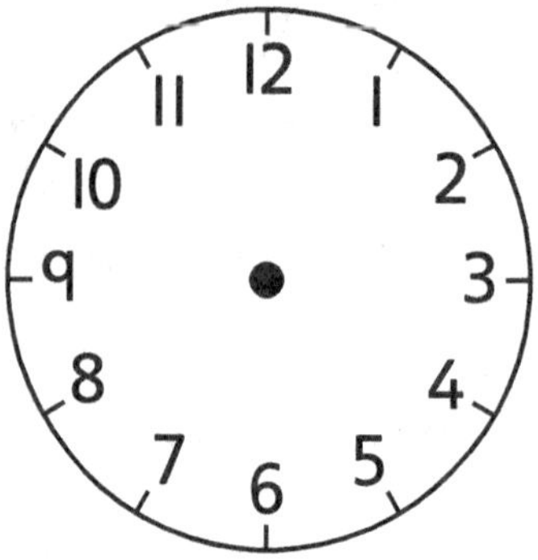

21. 4:30

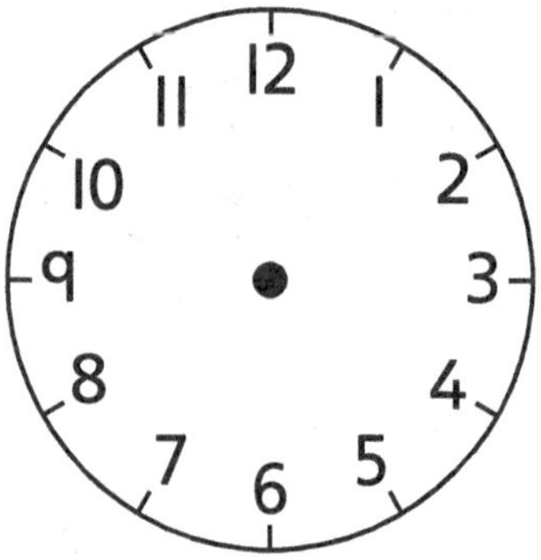

22. 3:30

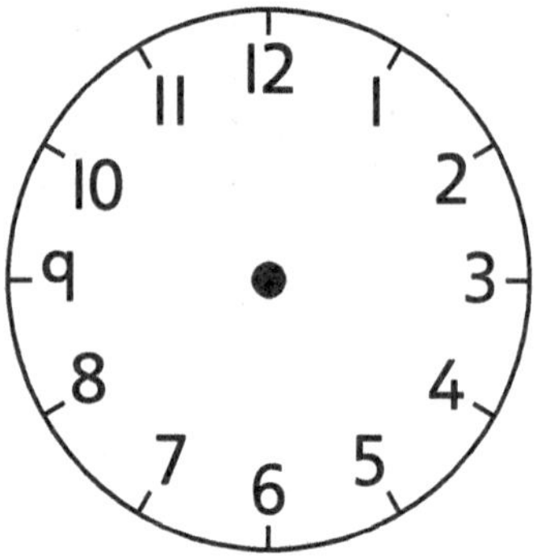

23. 10:30

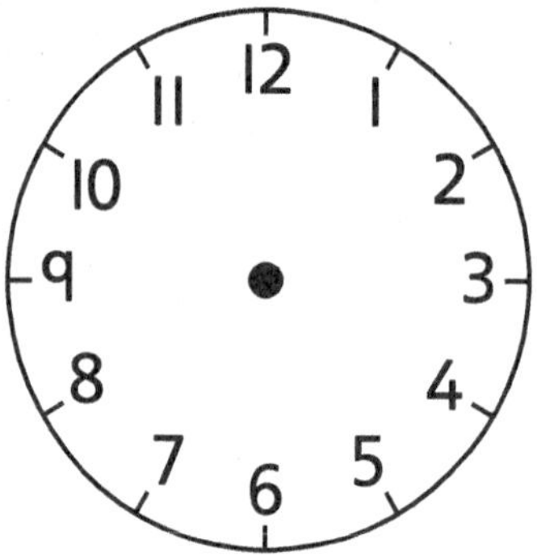

24. 8:30

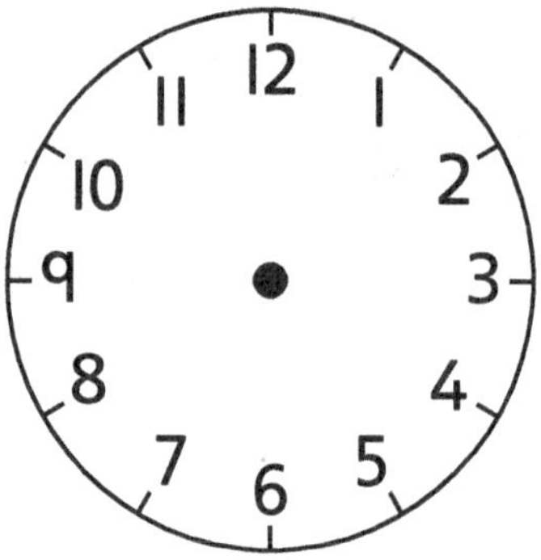

25. 7:30

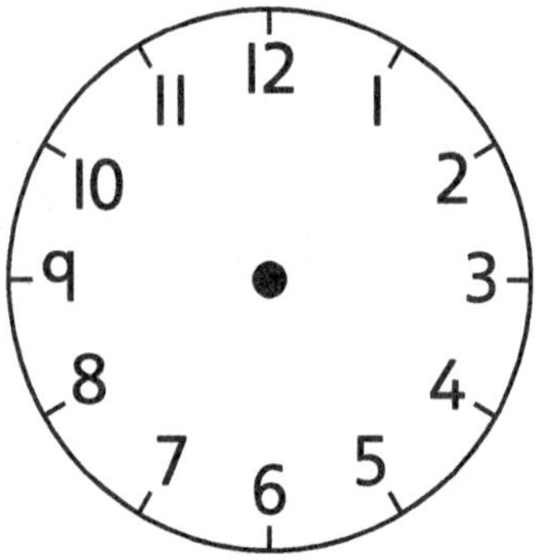

26. 1:30

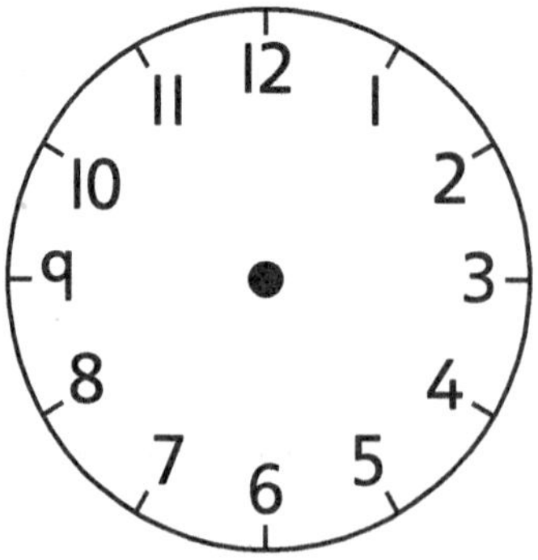

27. 6:30

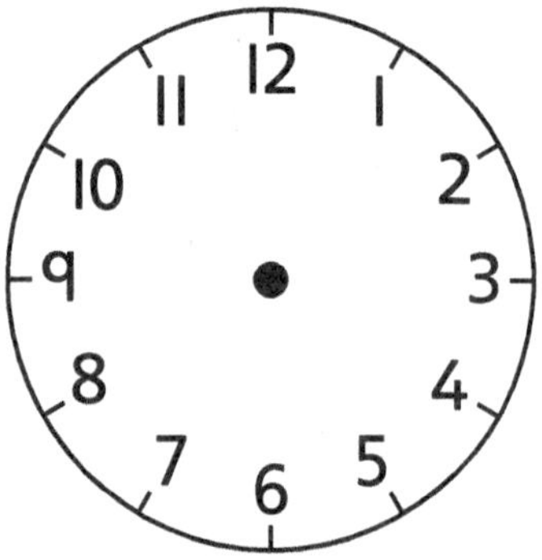

☐ Match the **digital clocks** to the **analog clocks**.

28.

1:00

4:00

11:00

29.

5:30

12:30

7:30

30.

1:30 | 7:00 | 2:00 | 8:30

MDI-17 Area

- ☐ Trace the **shapes** onto a piece of paper.
- ☐ Cut them out and compare the **areas**.
- ☐ Write **bigger** and **smaller** under each pair.

1.

______________________ ______________________

2.

______________________ ______________________

3.

______________________ ______________________

MDI-18 Capacity

☐ Write **less than, more than,** or **the same as.**

1.

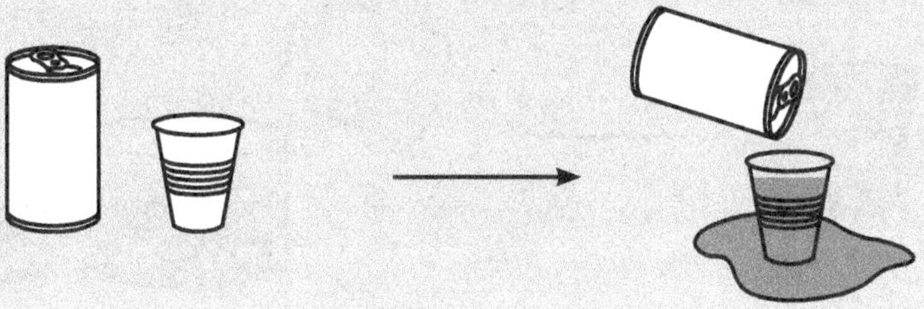

The  holds *more than* the .

2.

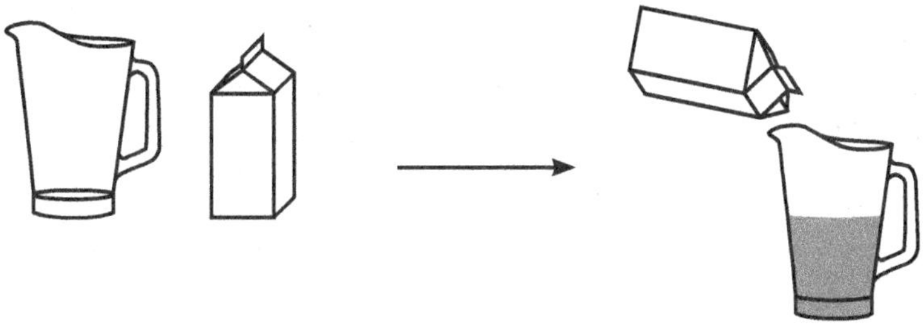

The  holds ____________________ the .

3.

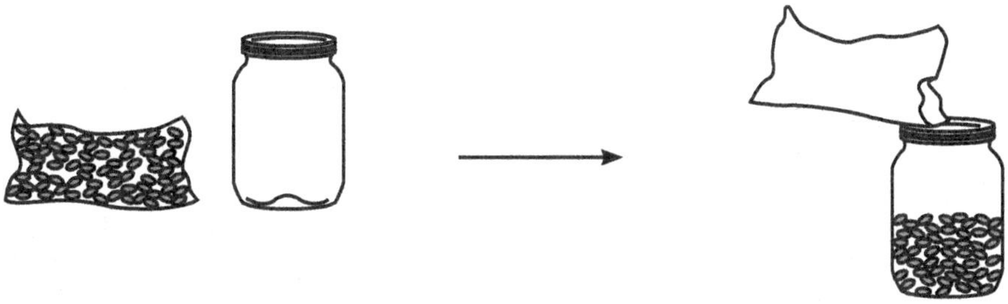

The  holds ____________________ the .

4.

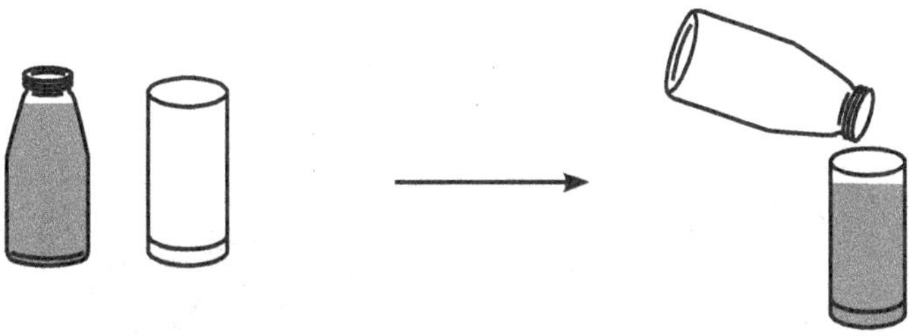

The  holds ____________________ the .

☐ Write **less than**, **more than**, or **the same as**.

5.

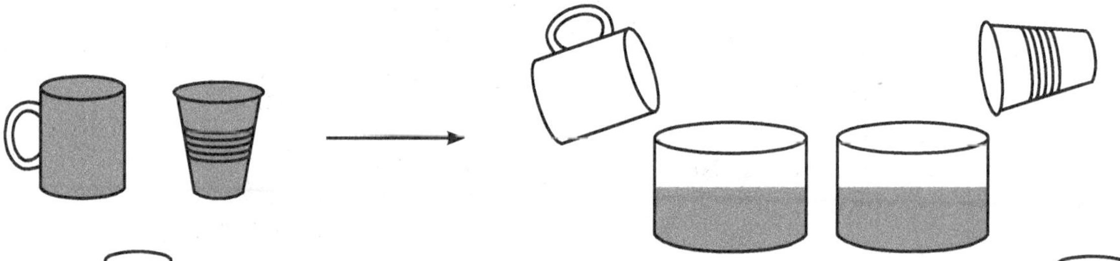

The  holds ______________________ the .

6.

The  holds ______________________ the 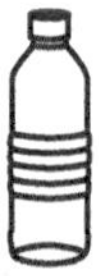.

7.

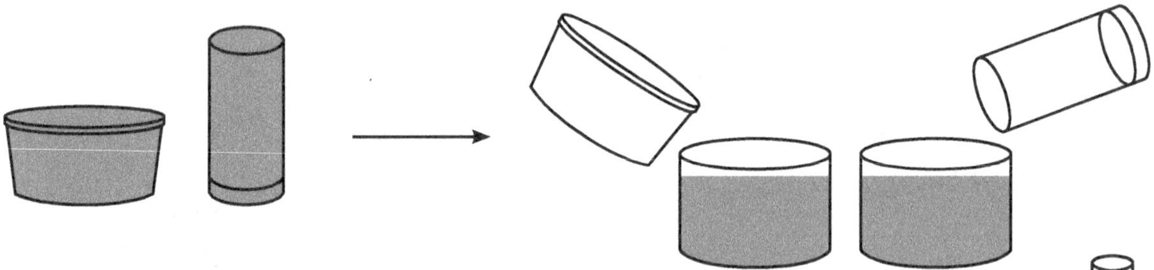

The  holds ______________________ the .

8.

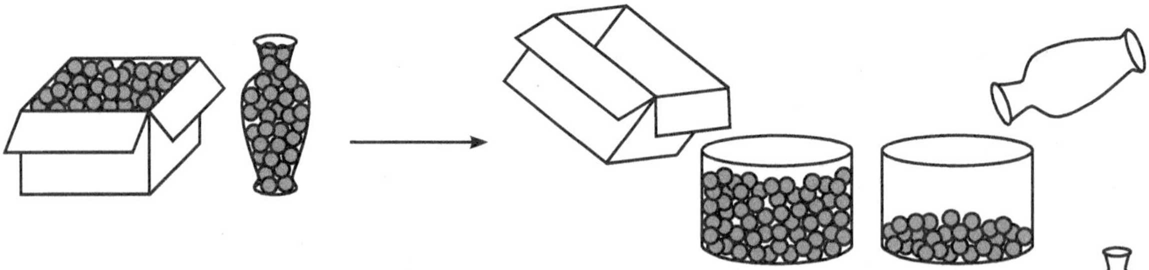

The 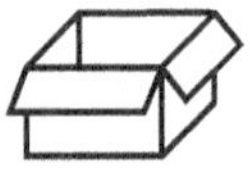 holds ______________________ the .

GI-I Sides and Vertices

open

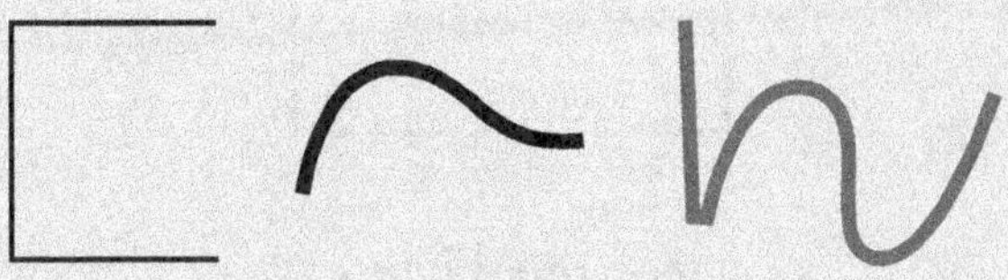

closed

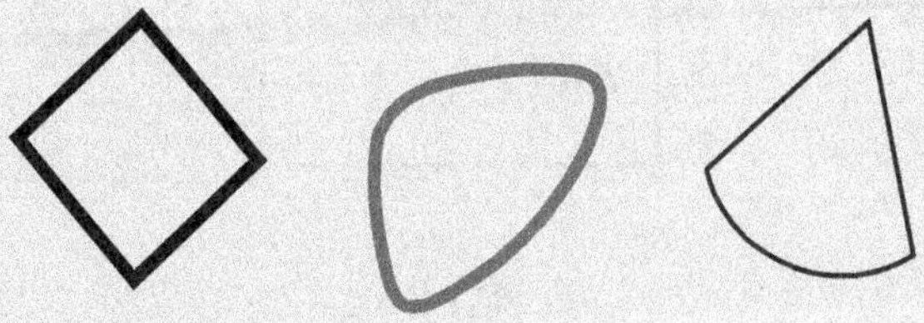

☐ Cross out the open lines.

☐ Circle the closed lines.

1.

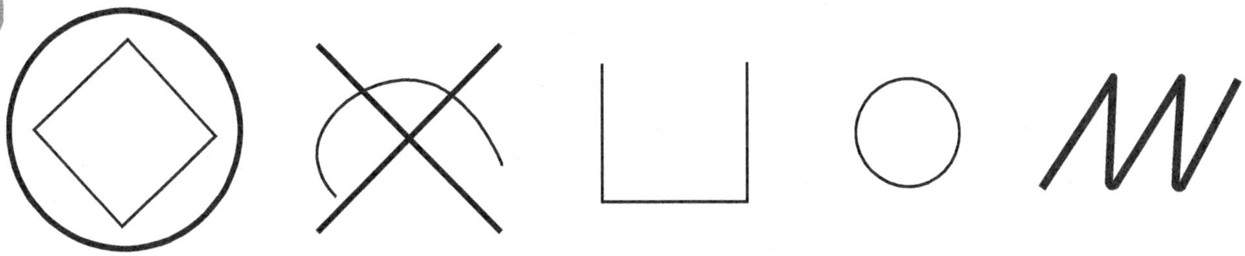

2.

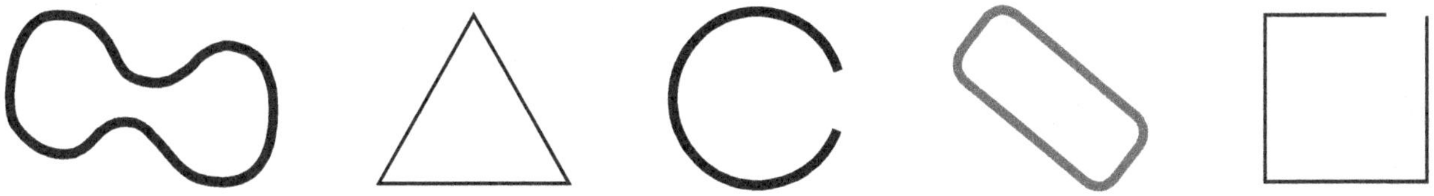

3.

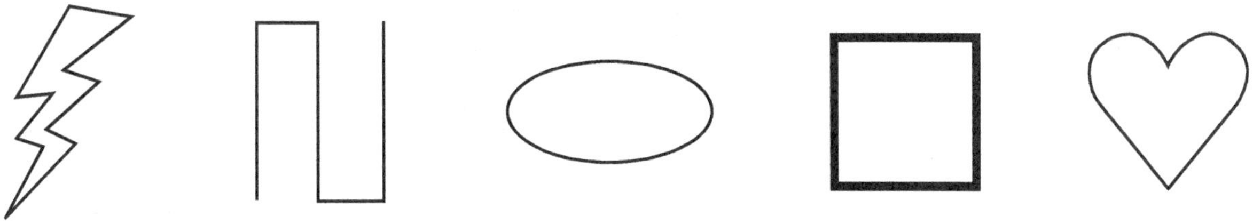

4.

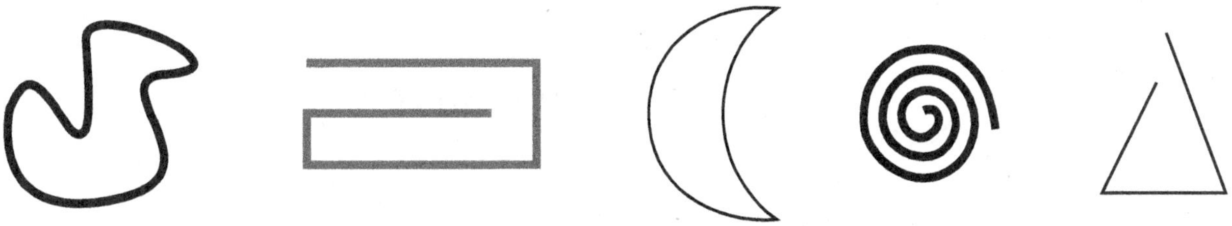

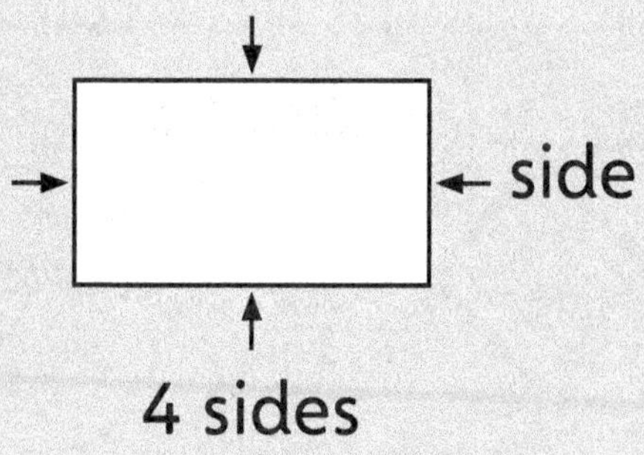

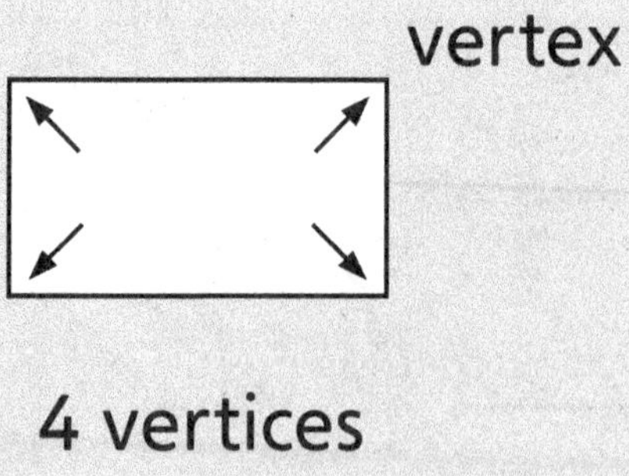

☐ Count the sides.

5.

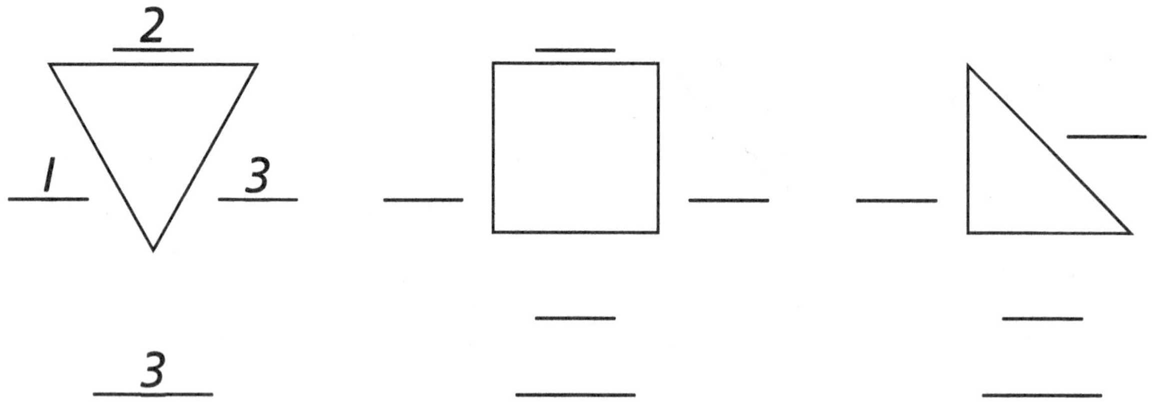

☐ Count the vertices.

6.

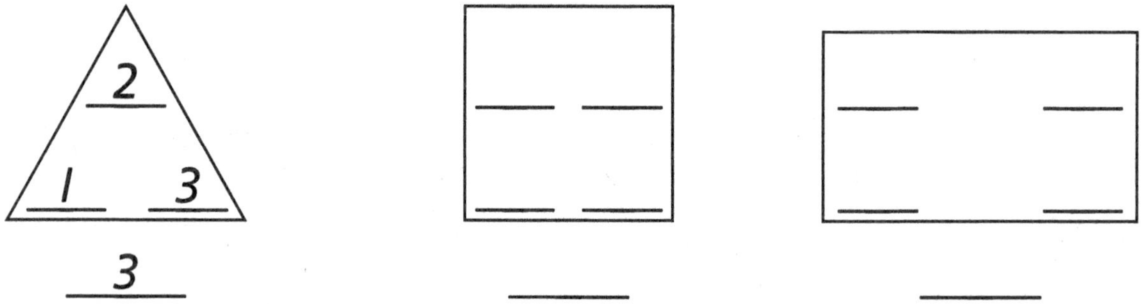

☐ Look at the shape and answer the questions.

7.

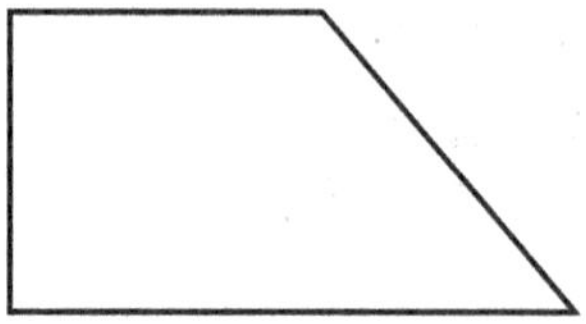

How many sides? ______

How many vertices? ______

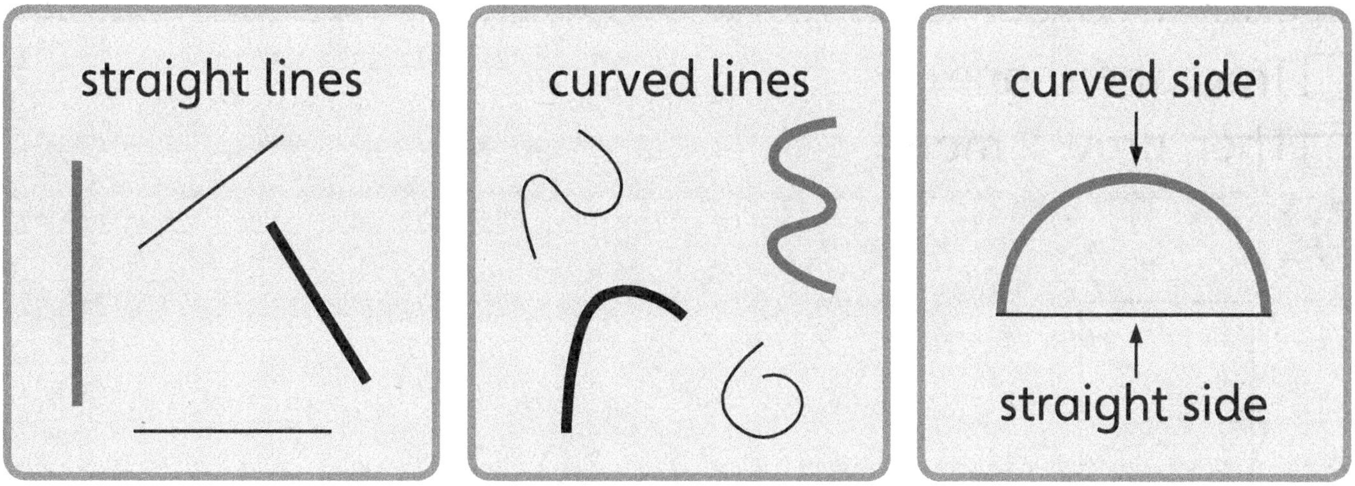

- [ ] Draw X on all the shapes that have a **straight** side.
- [ ] Draw O on all the shapes that have a **curved** side.

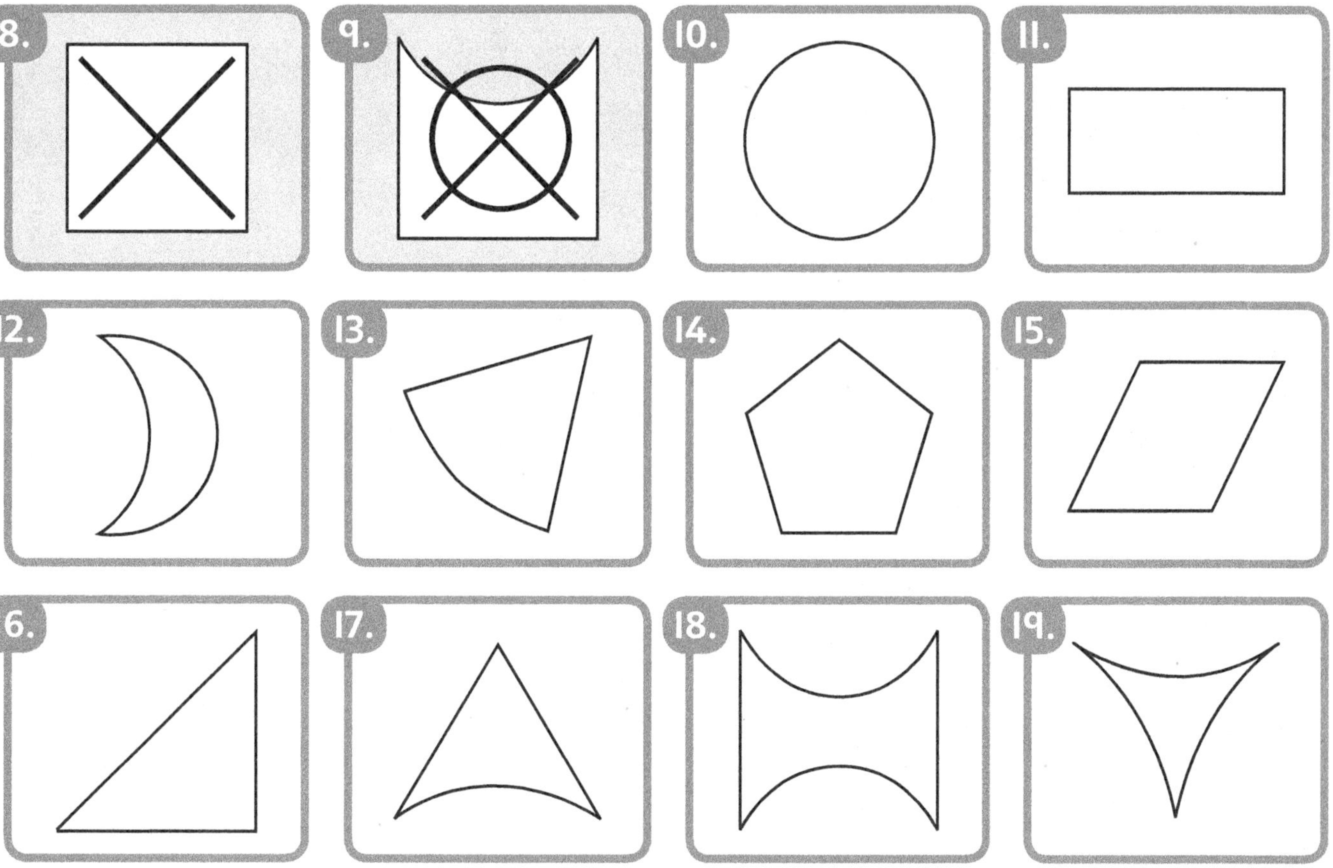

- [ ] Color the boxes of the shapes with X and O.

20. What letter do you see? __________

GI-2 More Sides and Vertices

☐ Trace the shape.

☐ Then draw 2 more.

1.

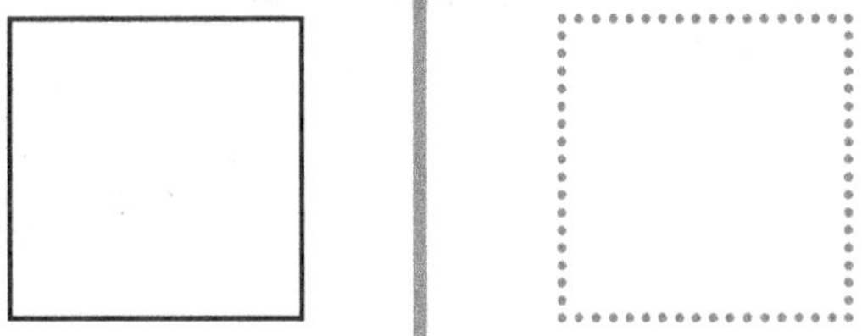

2.

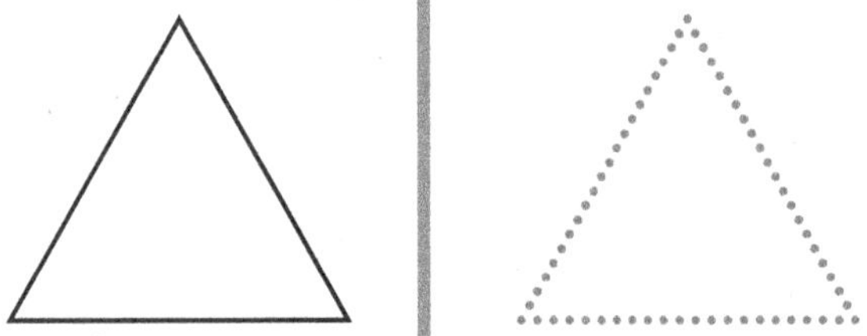

3.

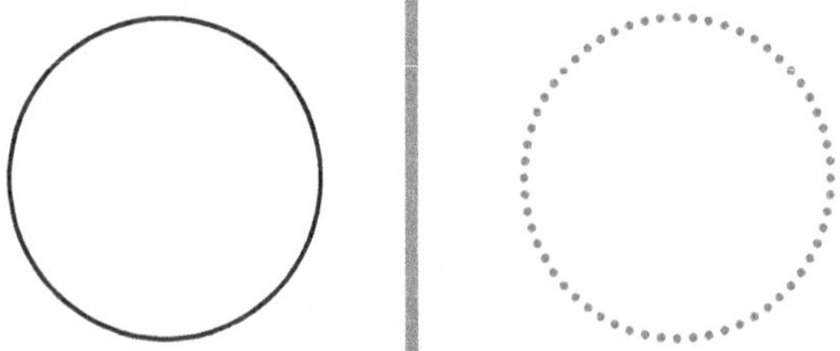

4.

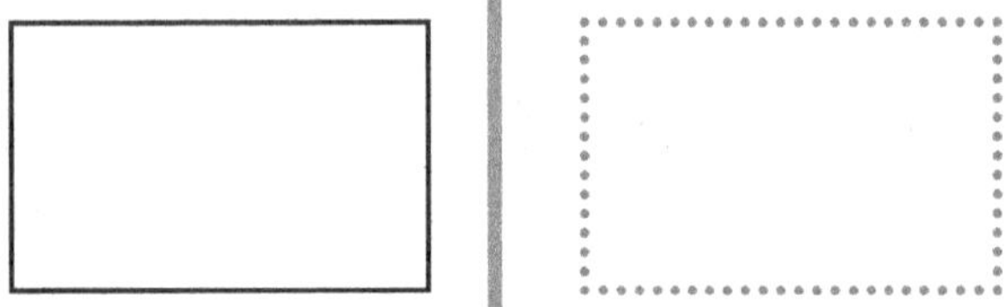

☐ Write ✓ if true.

☐ Write ✗ if not true.

5.

☑ 4 sides

☒ 3 vertices

6.

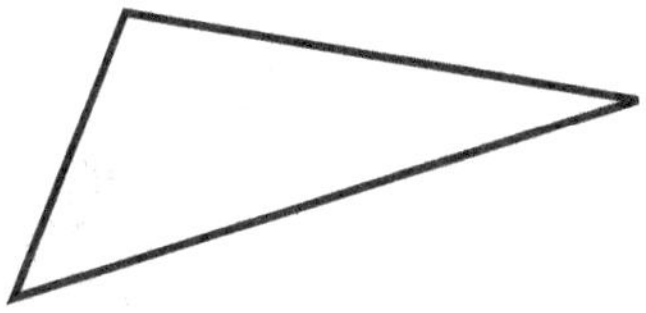

☐ 4 sides

☐ 4 vertices

7.

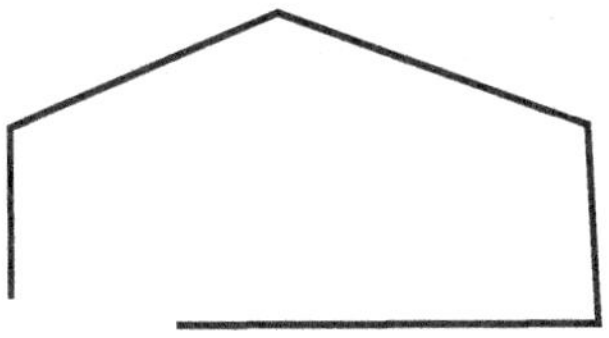

☐ 4 sides

☐ 4 vertices

8.

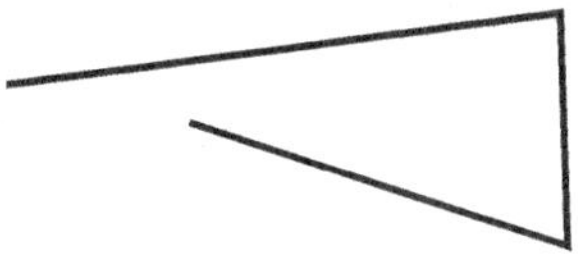

☐ 4 sides

☐ 4 vertices

☐ open

9.

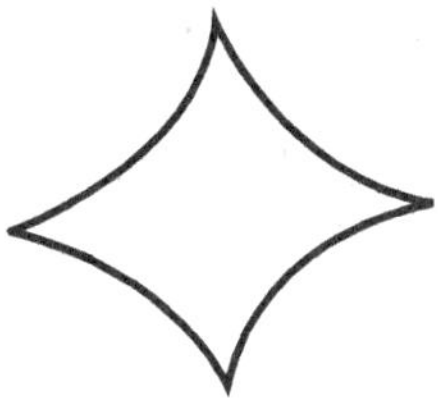

☐ 4 sides

☐ 4 vertices

☐ closed

10.

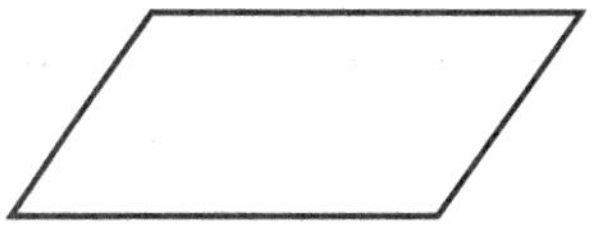

☐ 4 sides

☐ 4 vertices

☐ all sides curved

11.

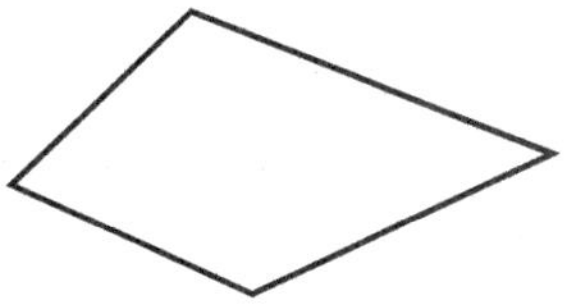

☐ 4 sides

☐ 4 vertices

☐ all sides straight

12.

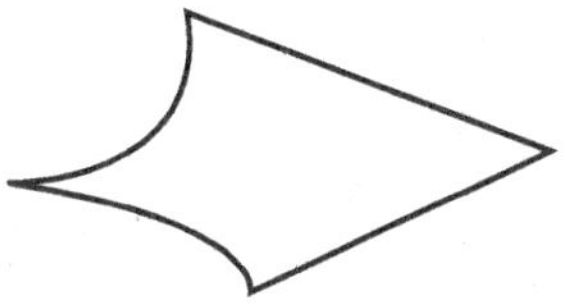

☐ 4 sides

☐ 4 vertices

☐ all sides straight

13.

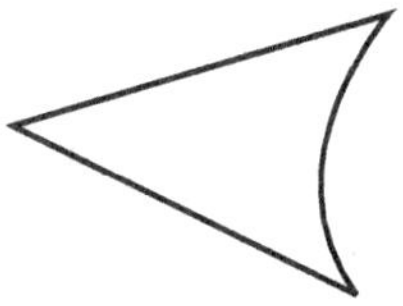

☐ 3 sides

☐ 3 vertices

☐ all sides curved

GI-3 Squares and Rectangles

squares

not squares

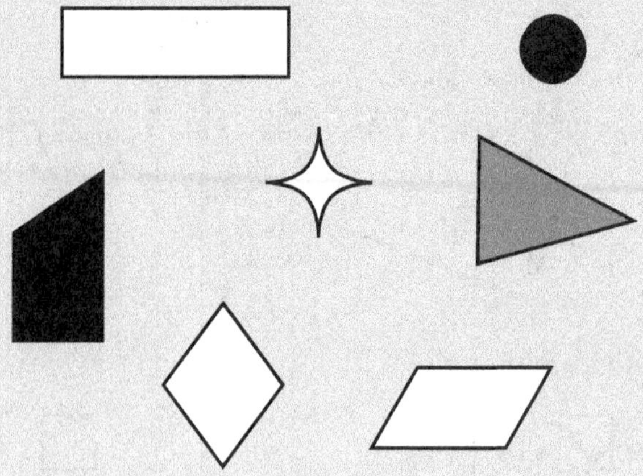

☐ Draw X on the shapes that are **not** squares.

1.

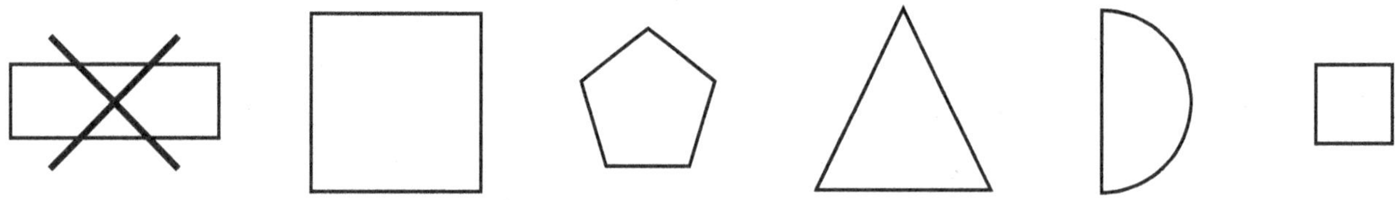

2.

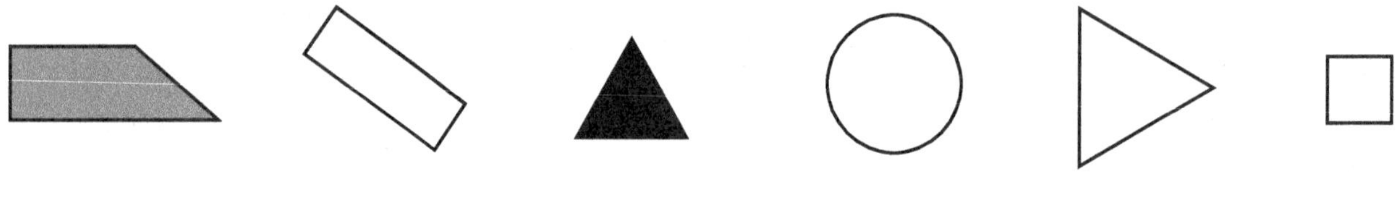

3.

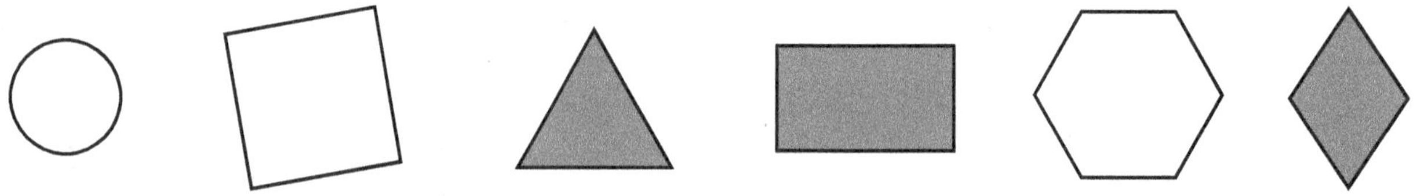

4.

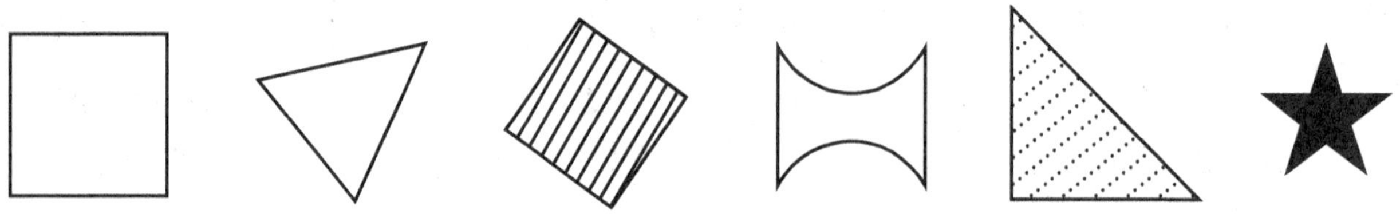

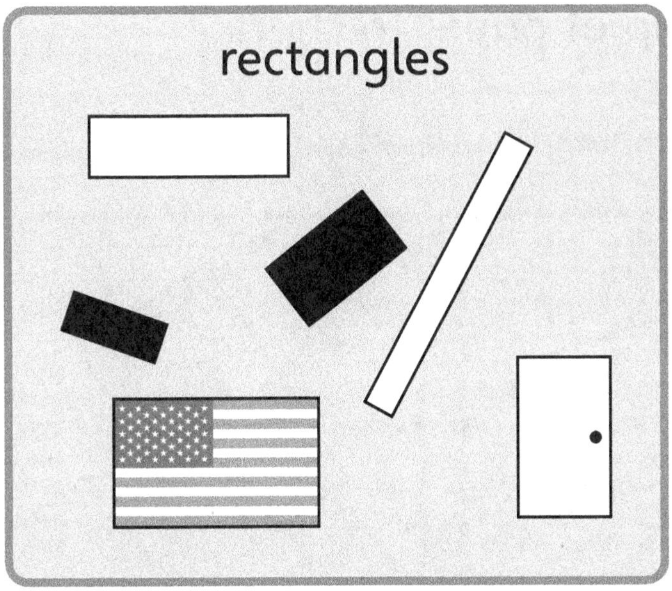

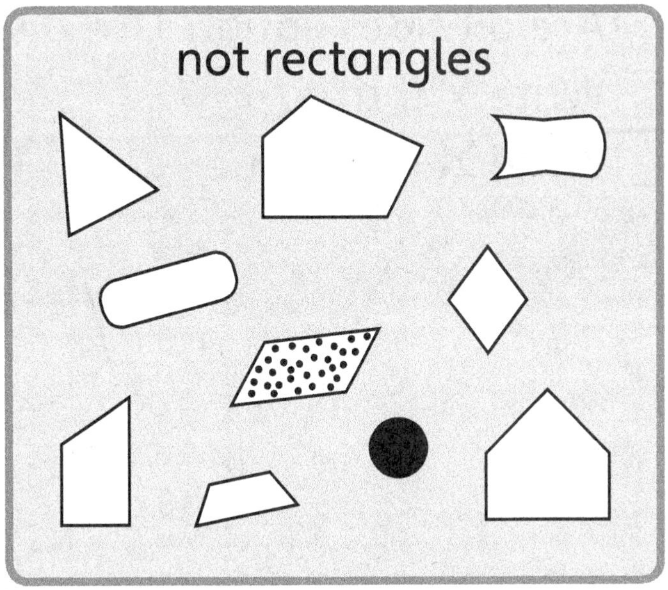

☐ Write ✓ under the rectangles.

5.

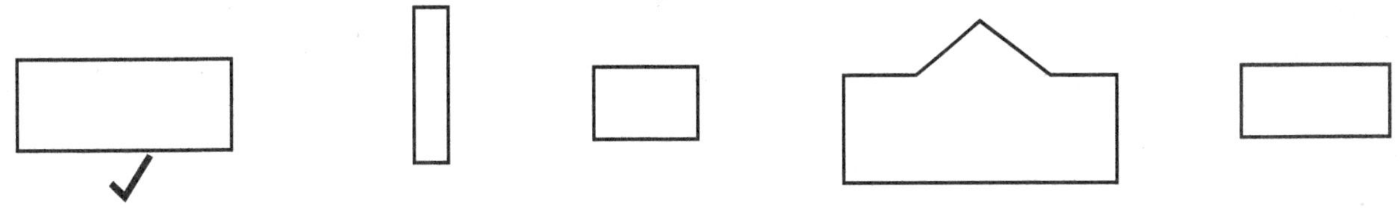

6.

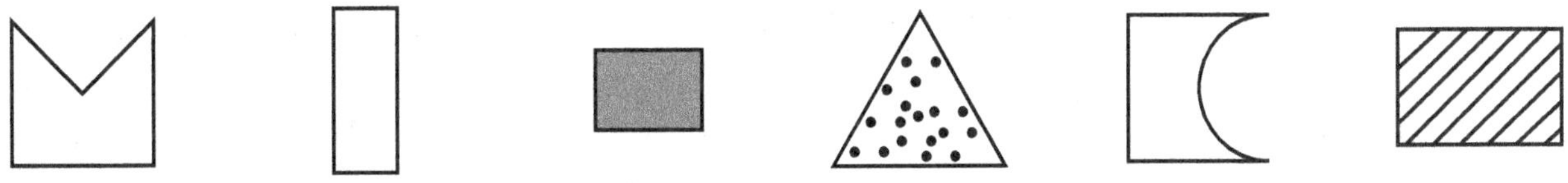

7.

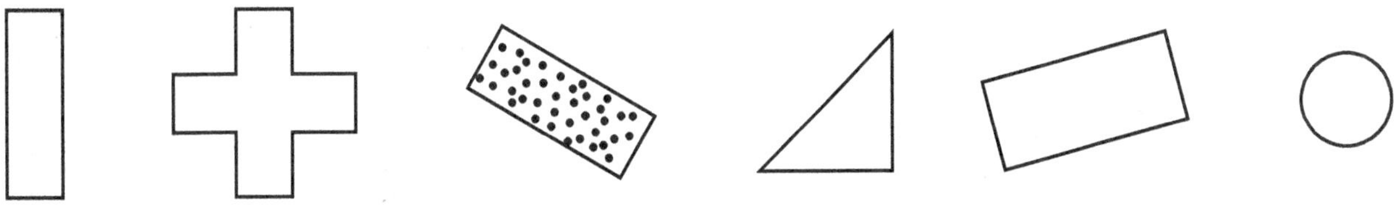

8.

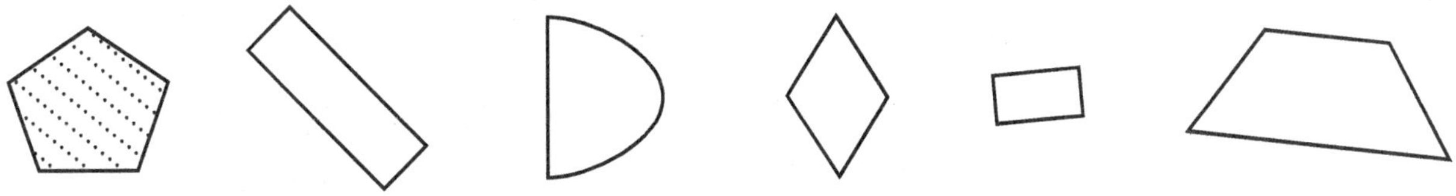

☐ Are all sides equal? Use a strip of paper to check.

☐ Write ✓ if true.

☐ Write ✗ if not true.

9.

☑ all sides equal

☑ 4 sides

☑ 4 vertices

☑ square

10.

☐ all sides equal

☐ 4 sides

☐ 4 vertices

☐ square

11.

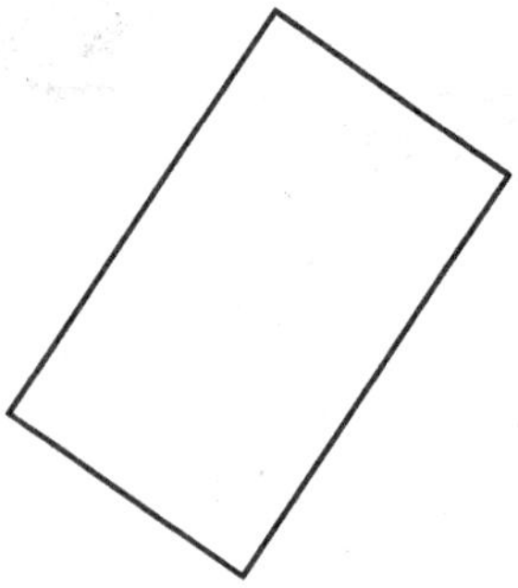

☐ all sides equal

☐ 4 sides

☐ 4 vertices

☐ rectangle

☐ Draw a shape with 4 sides.

12. Square

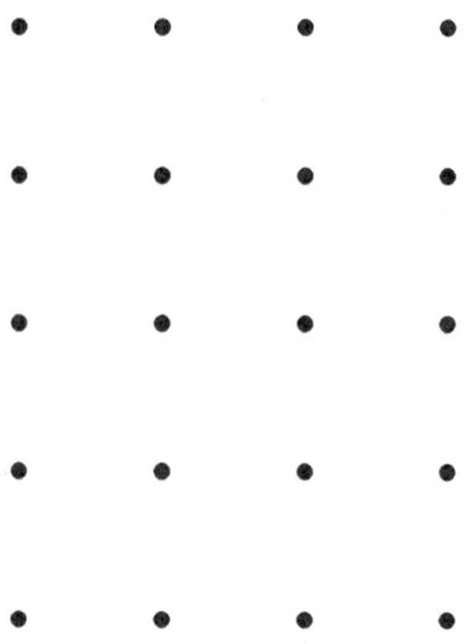

13. Rectangle but not a square

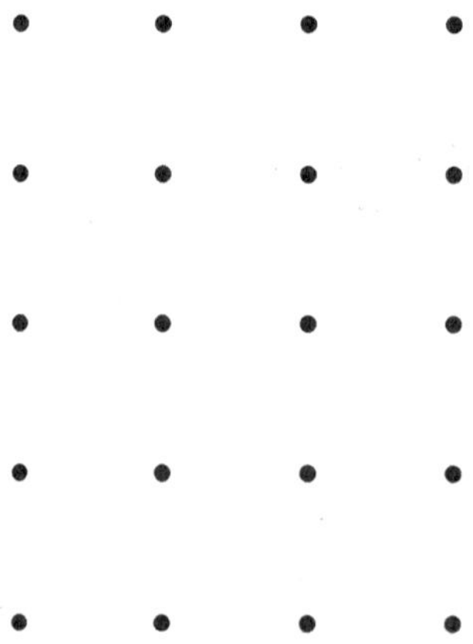

14. Not a square or a rectangle

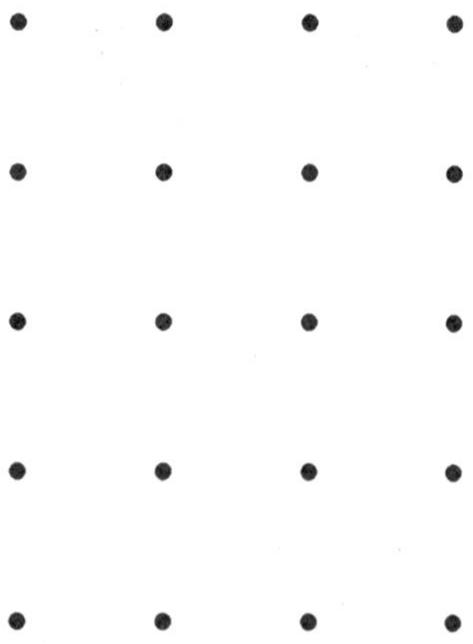

GI-4 Square Corners

Vertices like those of a square are called **square corners**.

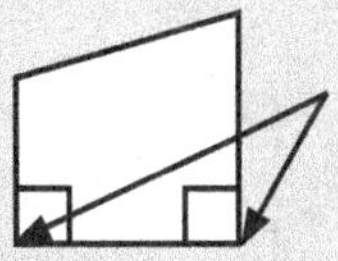

☐ Draw a small square in the square corners.

☐ How many vertices?

☐ How many square corners?

1.

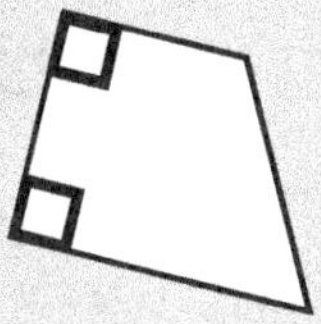

4 vertices

2 square corners

2.

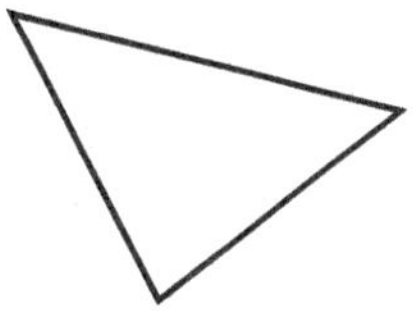

_____ vertices

_____ square corners

3.

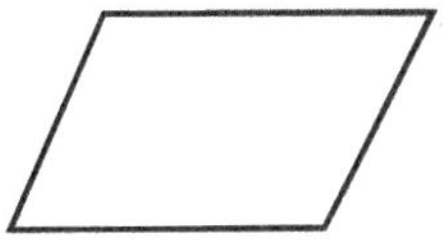

_____ vertices

_____ square corners

4.

_____ vertices

_____ square corners

5.

_____ vertices

_____ square corners

6.

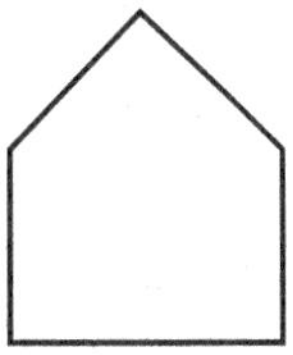

_____ vertices

_____ square corners

☐ Fill in the blanks.

7.

A  has _____ sides and _____ square corners.

A  has _____ sides and _____ square corners.

all sides equal

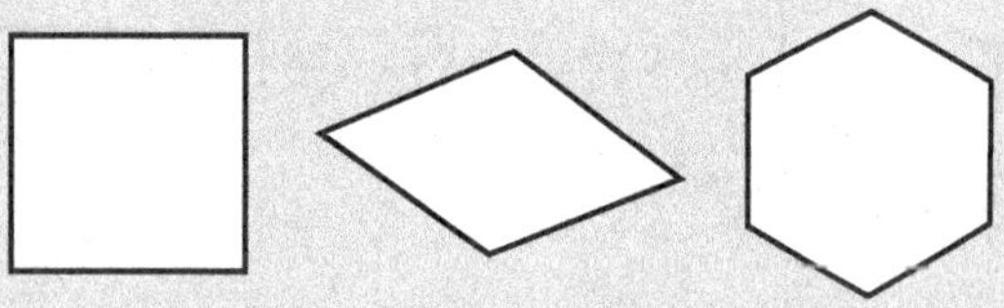

2 short sides, 2 long sides

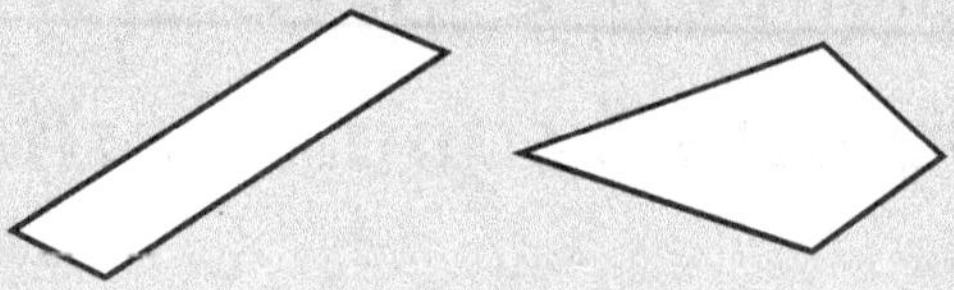

☐ Write ✓ if true.

☐ Write ✗ if not true.

8.

☑ all sides equal
☑ all square corners
☑ square

9.

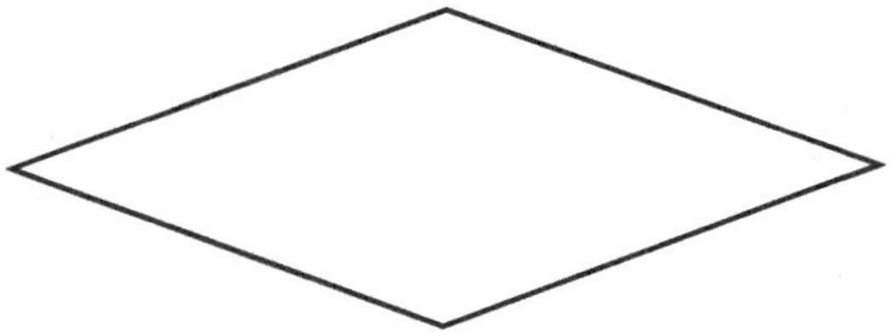

☐ all sides equal
☐ all square corners
☐ square

10.

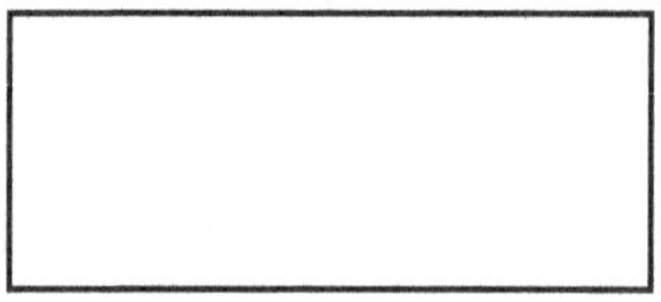

☐ all square corners
☐ 2 short sides, 2 long sides
☐ rectangle

11.

☐ all square corners
☐ 2 short sides, 2 long sides
☐ rectangle

12. Is this shape a square?
Is it a rectangle? Explain.

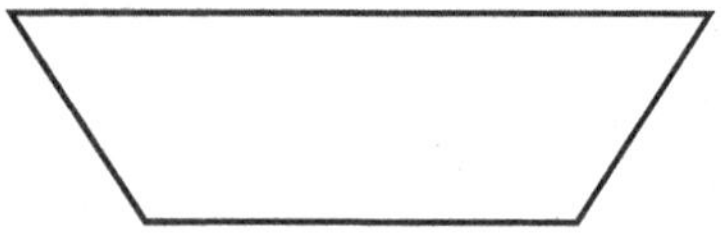

GI-5 Triangles

triangles

not triangles

☐ Circle the triangles.

1.

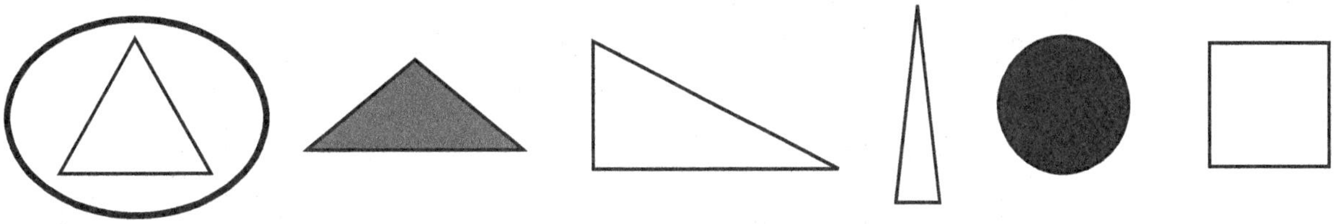

2.

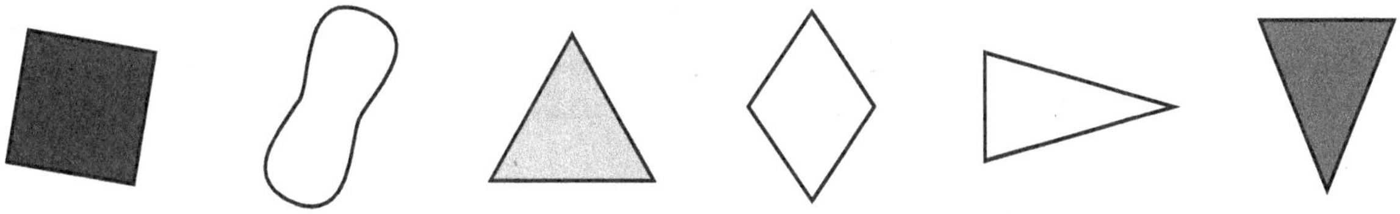

3.

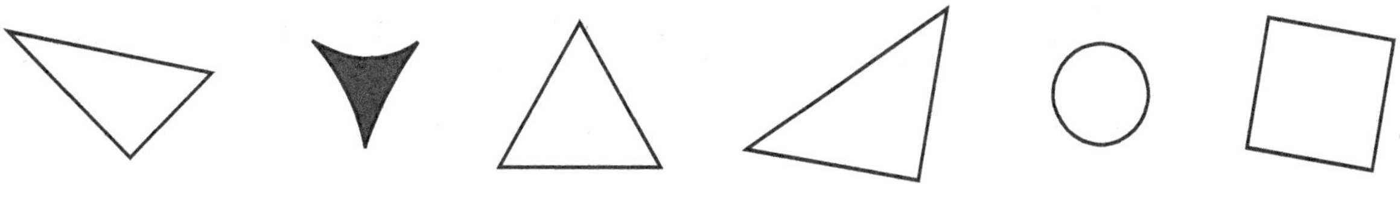

4.

☐ Write ✓ if true.

☐ Write ✗ if not true.

5.

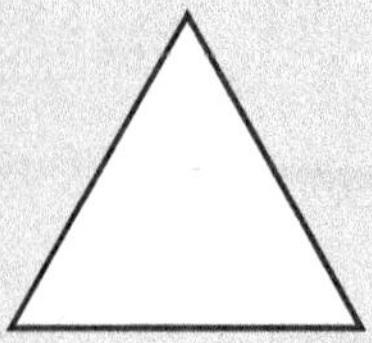

☑ 3 sides

☑ 3 vertices

6.

☐ 3 sides

☐ 3 vertices

7.

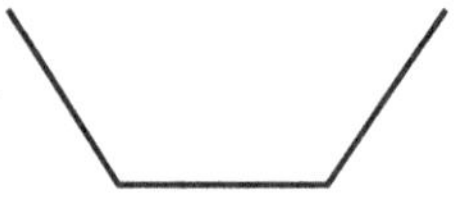

☐ 3 sides

☐ 3 vertices

8.

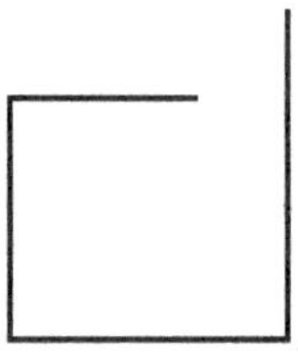

☐ 3 sides

☐ 3 vertices

9.

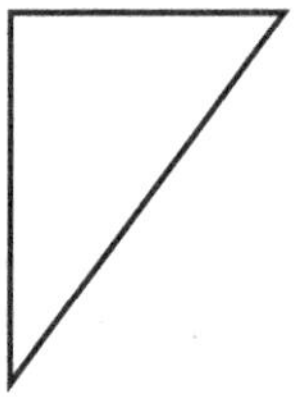

☐ 3 sides

☐ 3 vertices

☐ triangle

10.

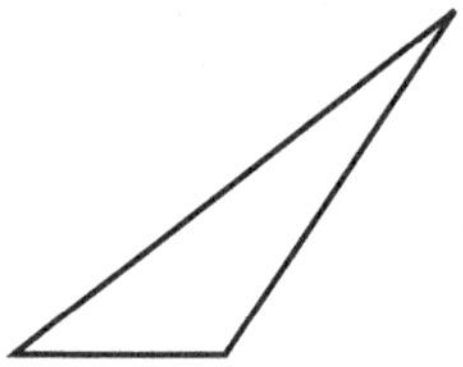

☐ 3 sides

☐ 3 vertices

☐ triangle

11.

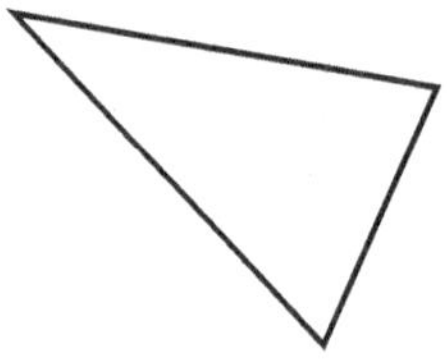

☐ 3 sides

☐ 3 vertices

☐ triangle

12.

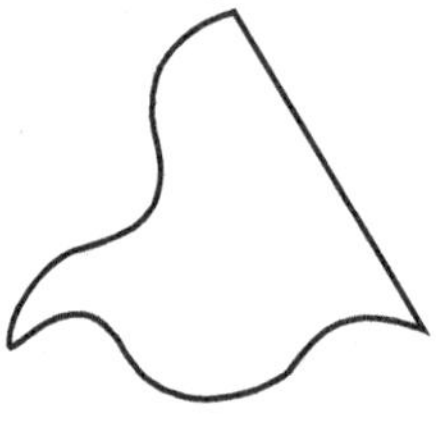

☐ 3 sides

☐ 3 vertices

☐ triangle

☐ Fill in the blanks.

13.

Triangles have _____ straight sides.

Triangles have _____ vertices.

☐ Draw the shape.

14. A shape with 3 straight sides.

Is it a triangle? ________

15. A shape with 3 sides that is not a triangle.

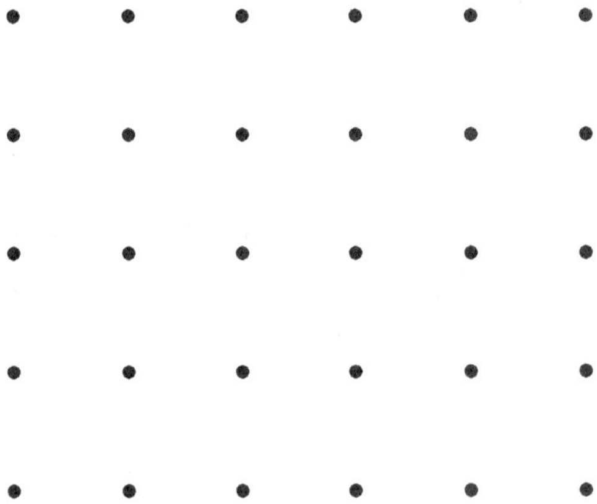

☐ Look at the shapes and answer the questions.

16.

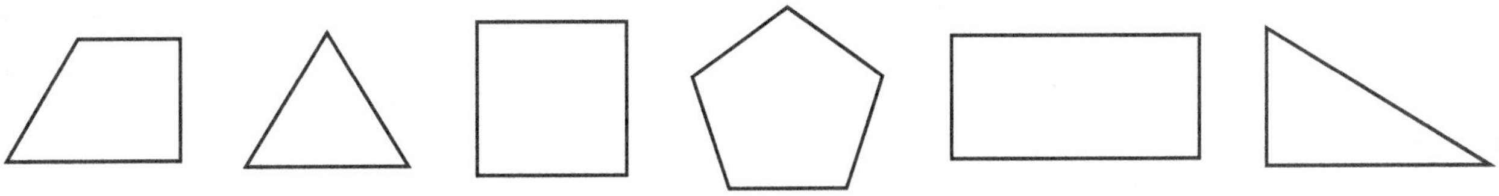

How many shapes have 4 vertices? _____

How many shapes have 3 sides? _____

17. BONUS

Draw a square, a rectangle, or a triangle.
How do you know you drew it correctly?

18. BONUS

Draw a triangle with a square corner.

GI-6 Creating Smaller Shapes

☐ Draw lines to make the big shape into smaller shapes.

1. 2 triangles

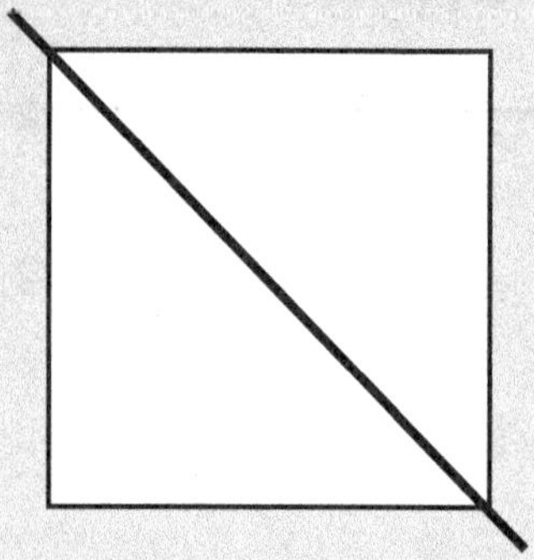

2. 3 triangles

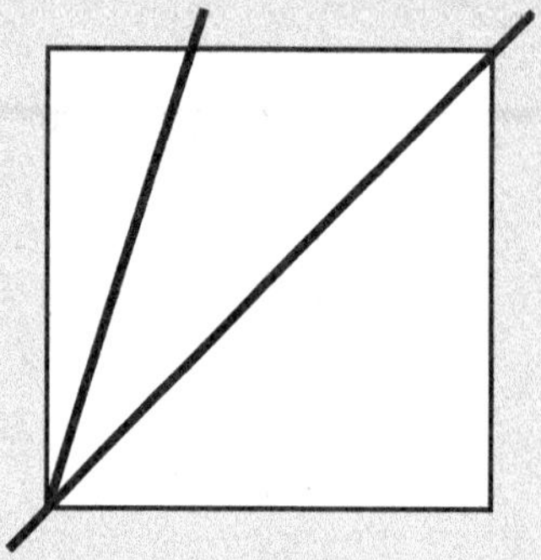

3. 4 triangles

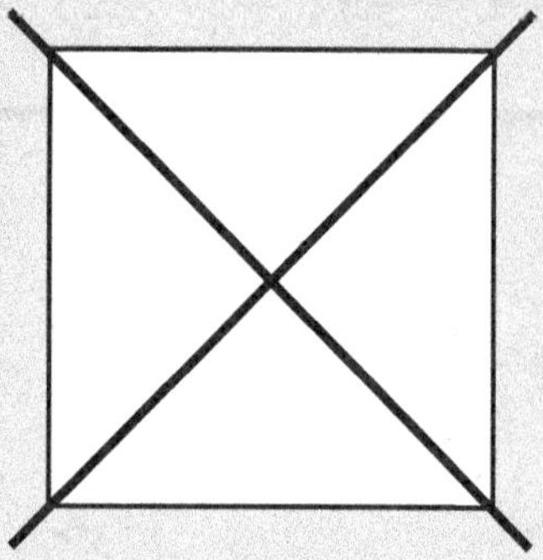

4. 2 rectangles

5. 4 squares

6. 4 rectangles

7. 3 rectangles

8. 4 triangles

9. 8 triangles

☐ What shapes can you make if you cut along the line?

10.

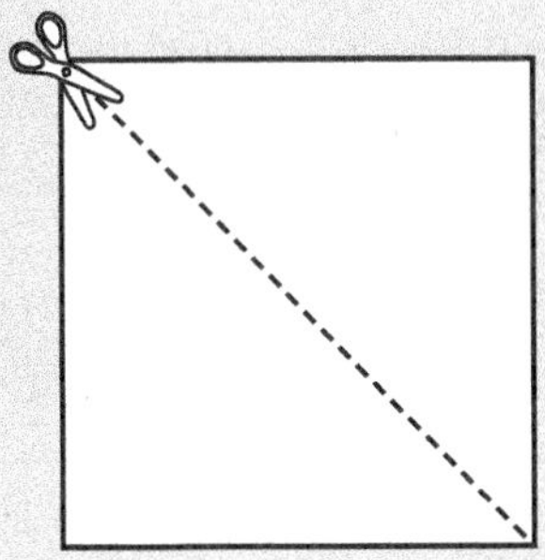

☑ triangle
☒ rectangle
☒ square

11.

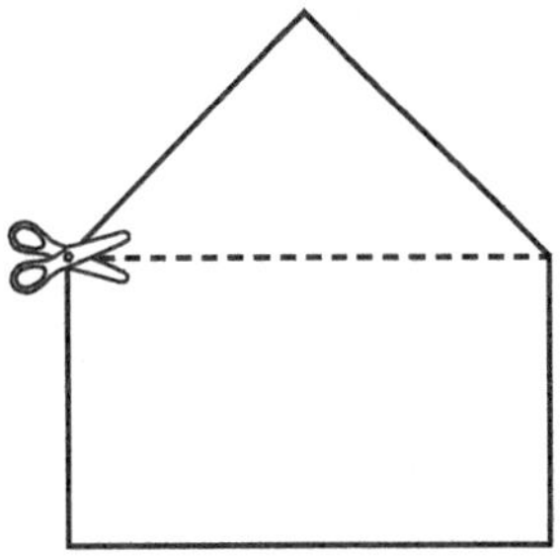

☐ triangle
☐ rectangle
☐ square

12.

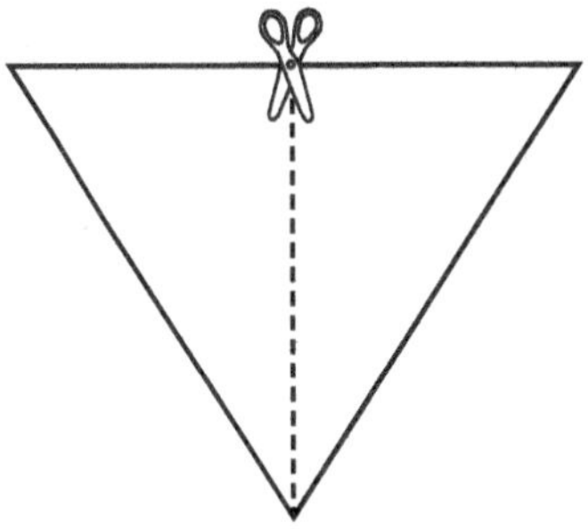

☐ triangle
☐ rectangle
☐ square

☐ What shapes can you make if you cut along both dotted lines?

13.

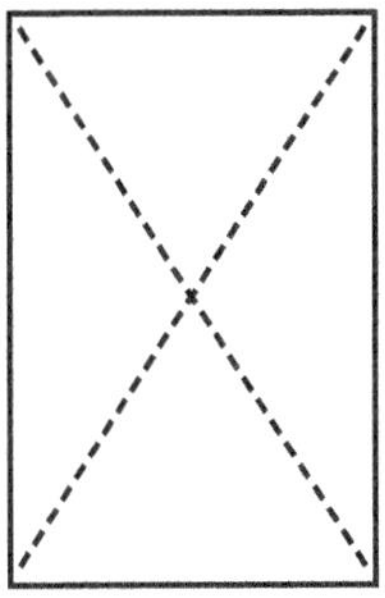

☐ triangle
☐ rectangle
☐ square

14.

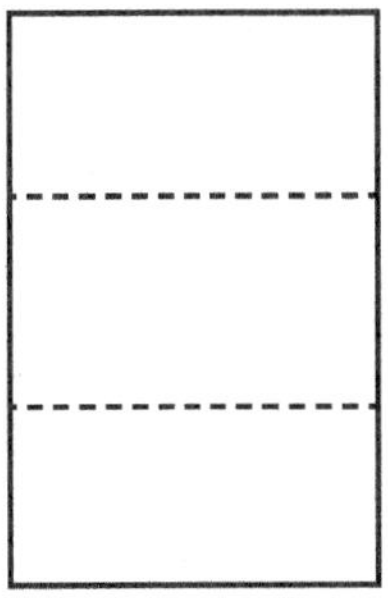

☐ triangle
☐ rectangle
☐ square

15.

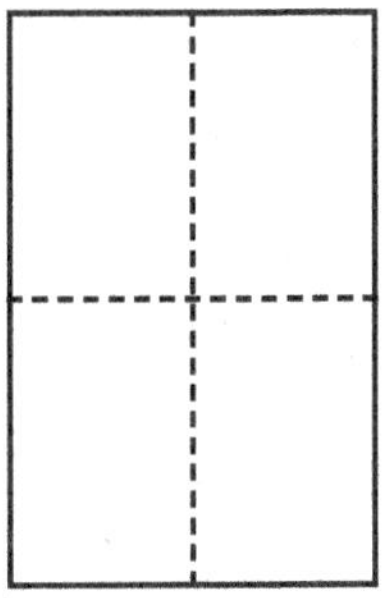

☐ triangle
☐ rectangle
☐ square

16. Cut a piece of paper to make 2 rectangles.

17. Cut a piece of paper to make 2 triangles.

GI-7 Pattern Blocks

☐ Circle the **pattern blocks** used to make the shape.

1.

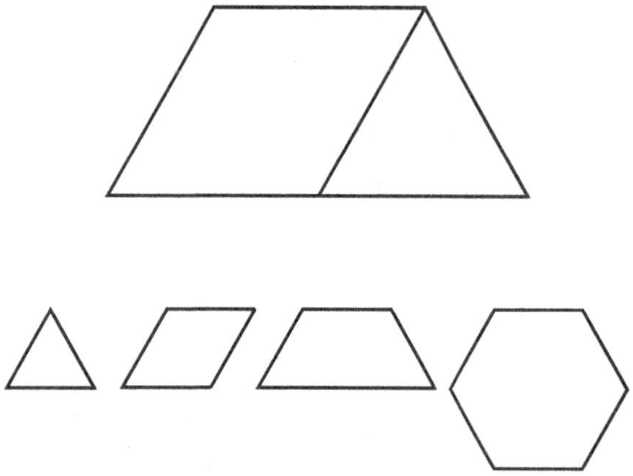

2.

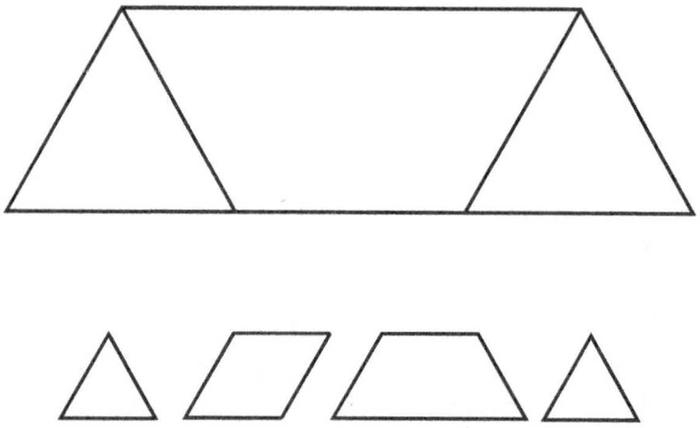

3.

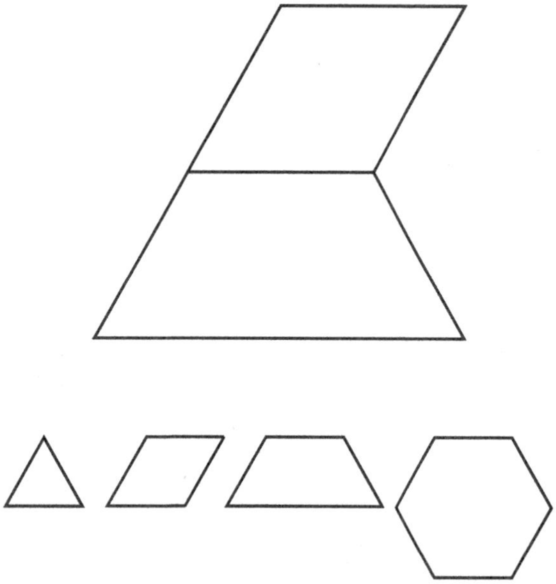

4.

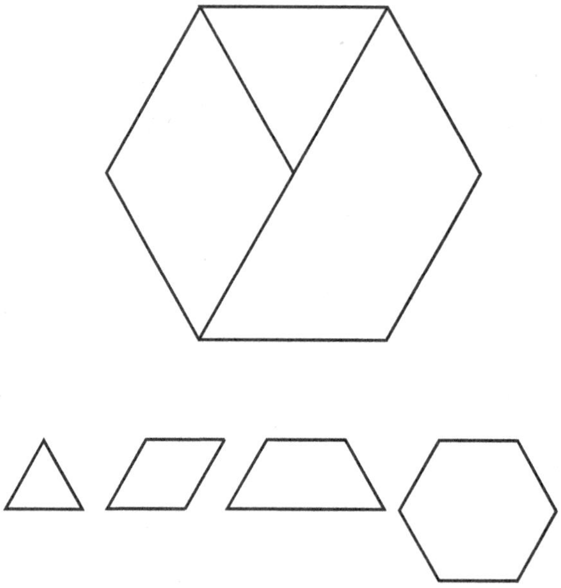

☐ Use 2 triangles to make the **rhombus**.

☐ Show how you did it.

5.

rhombus

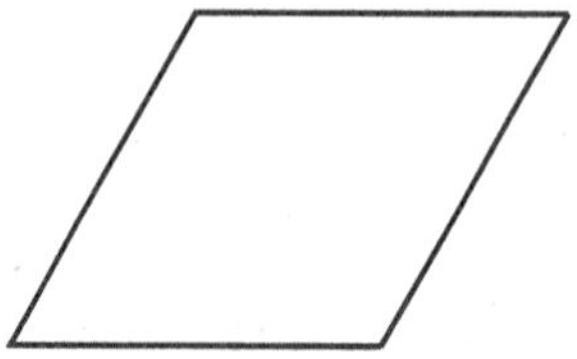

☐ Use 2 pattern blocks to make each shape.

☐ Circle the blocks you used.

6.

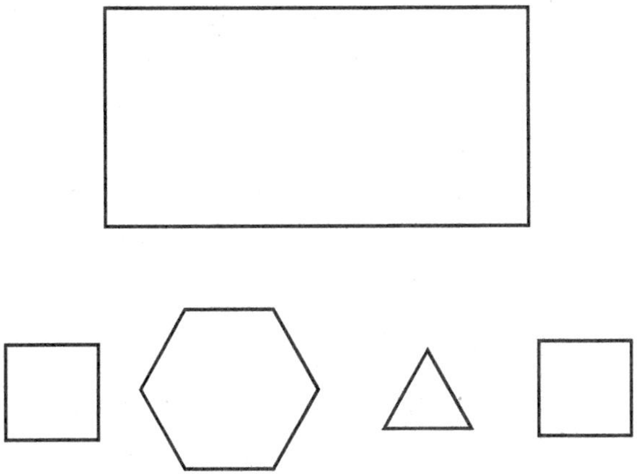

7.

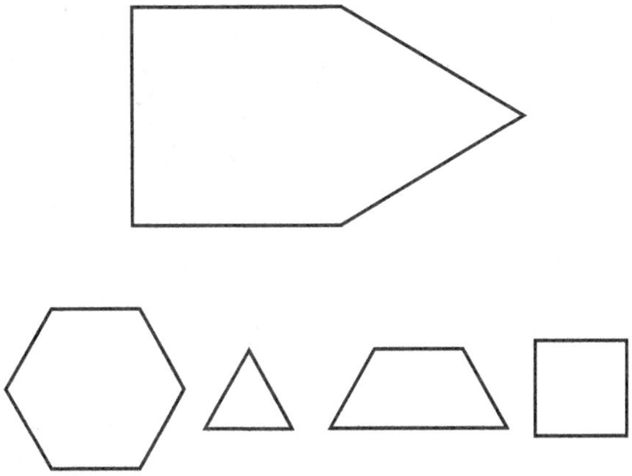

8.

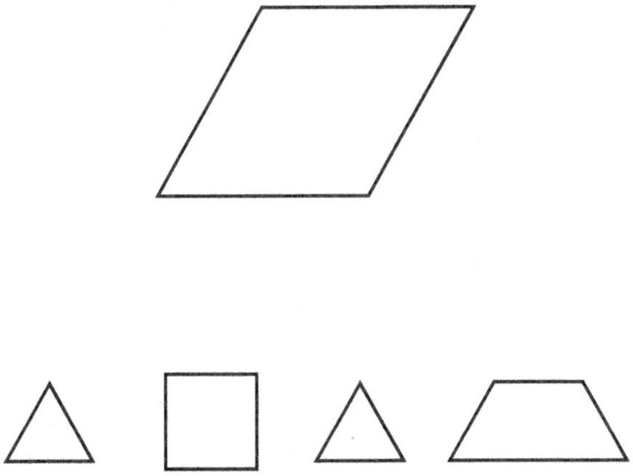

9. BONUS

Use 3 blocks.

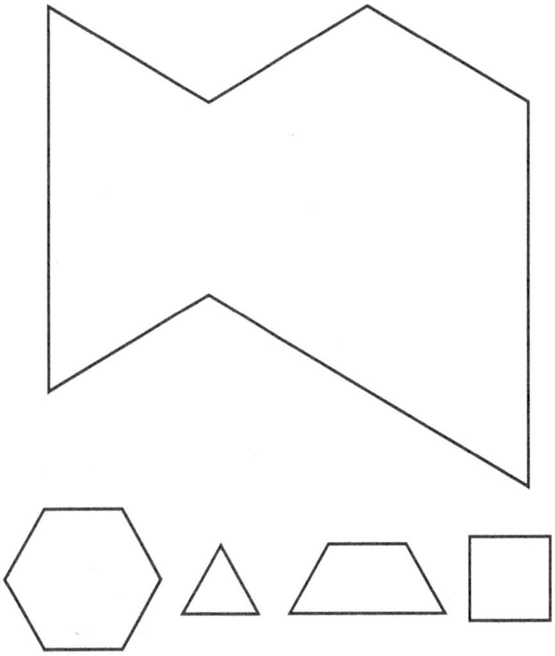

10.

Make a hexagon 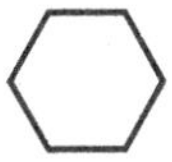

using 2 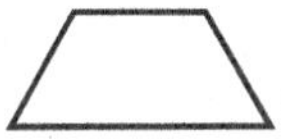.

11.

Make a hexagon 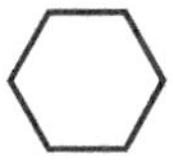 using 1 triangle, 1 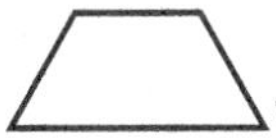,

and this rhombus 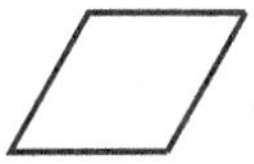.

GI-8 Circles

circles

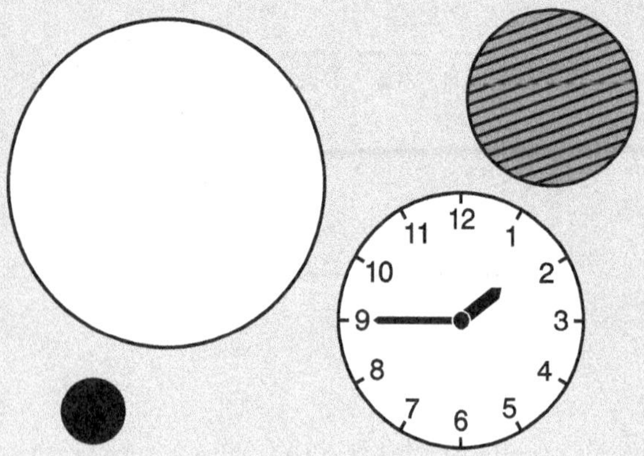

not circles

☐ Draw ✕ on the shapes that are **not** circles.

1.

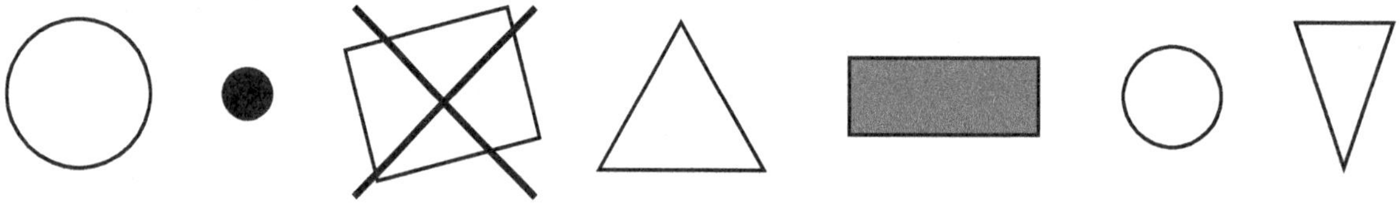

2.

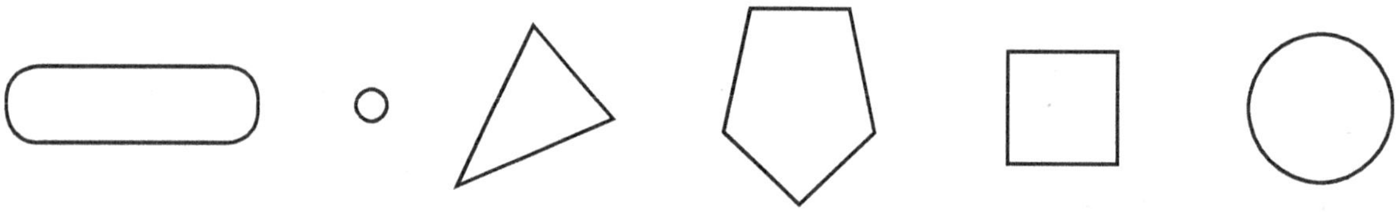

3.

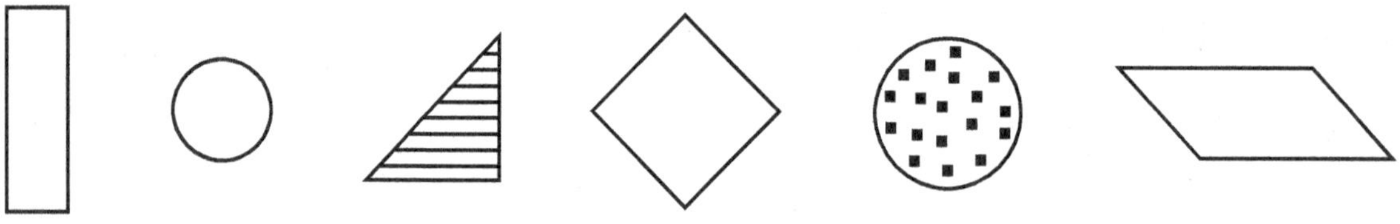

4.

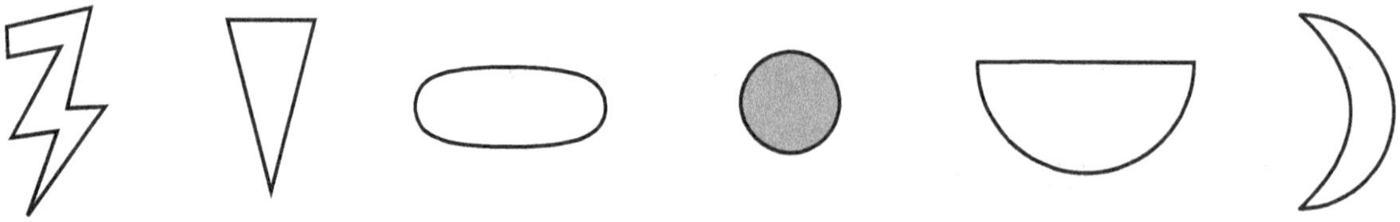

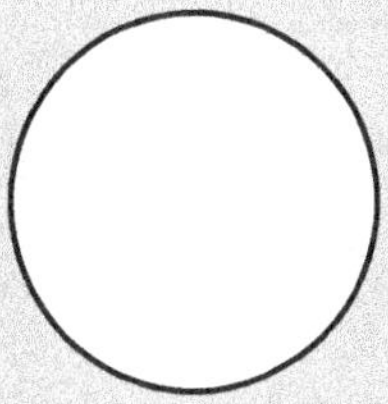

A circle is closed and round.

A circle does not have any straight sides or vertices.

☐ Write how many straight sides and vertices.

5.

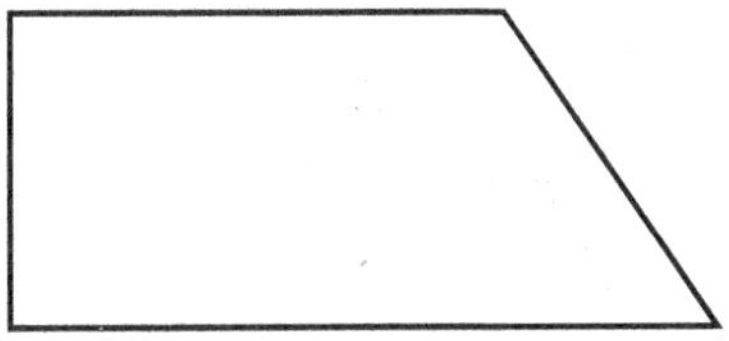

_____ straight sides

_____ vertices

6.

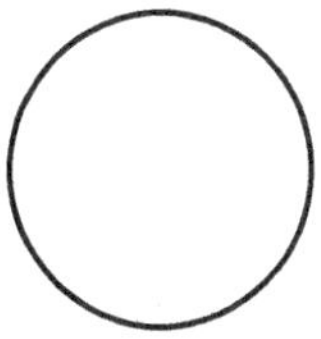

_____ straight sides

_____ vertices

☐ Is the line closed? Write ✓ or ✗.

7.

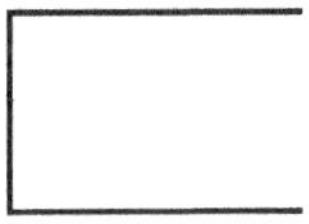

☐ closed

8.

☐ closed

9.

☐ closed

10.

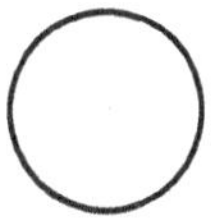

☐ closed

11.

☐ closed

12.

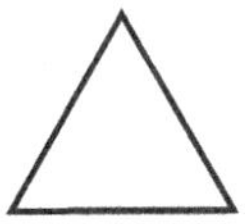

☐ closed

13. Find an object in your classroom that is a circle.

A circle can be cut into 2 equal parts.
Each part is called a **half circle**.

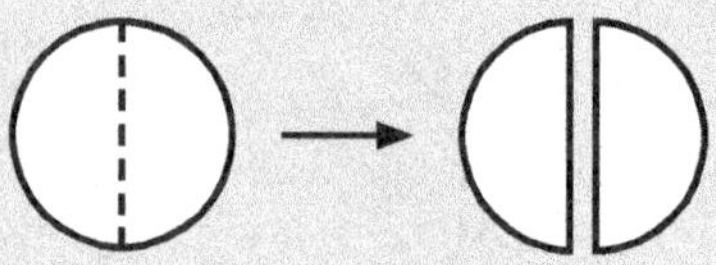

☐ Color the half circles.

14.

A circle can be cut into 4 equal parts.
Each part is called a **quarter circle**.

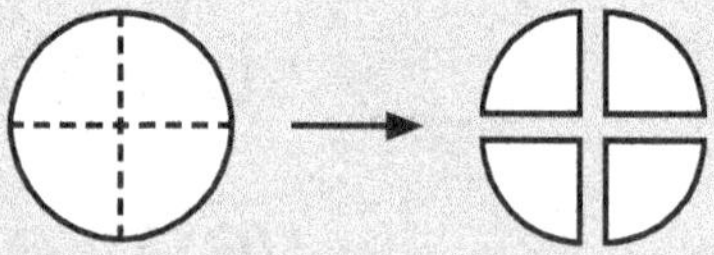

☐ Color the quarter circles.

15.

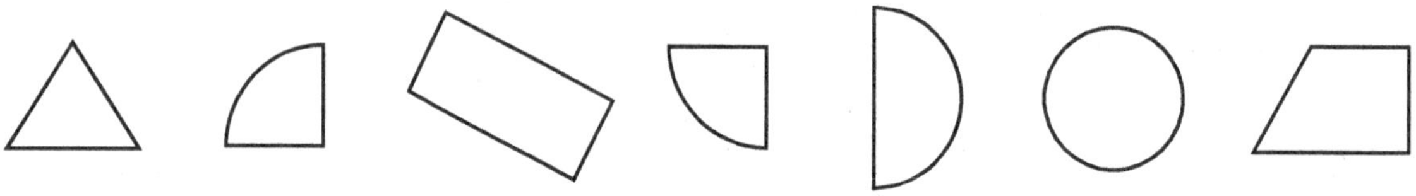

☐ Color the house.

16.

blue **half circles**

red **squares**

green **triangles**

trace around 2 **rectangles**

G1-9 Half

A **whole** can be cut into **equal parts**.
Equal parts are the same size.

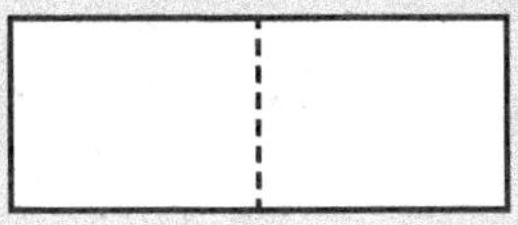

2 equal parts

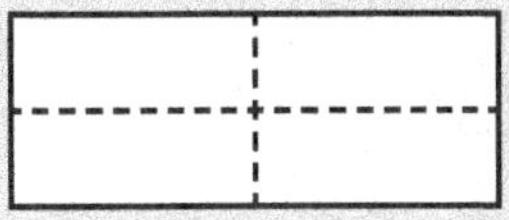

4 equal parts

☐ How many equal parts?

1.

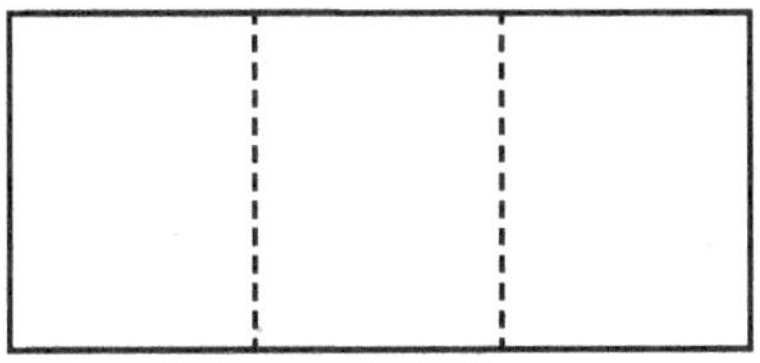

_____ equal parts

2.

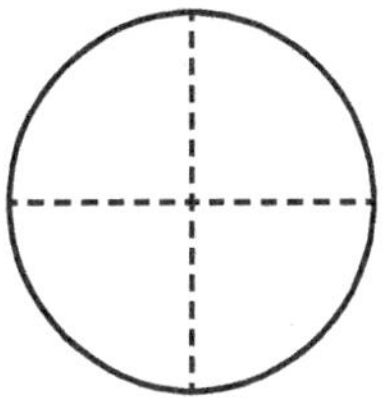

_____ equal parts

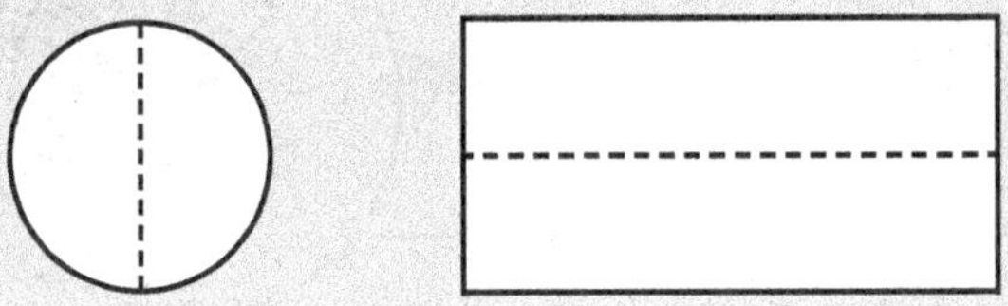

Each shape is cut into 2 equal parts.
The parts are called **halves**.
One part is called a **half**.

☐ How many?

3.

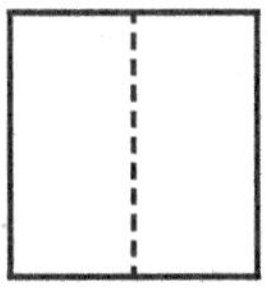

_____ equal parts

_____ halves

4.

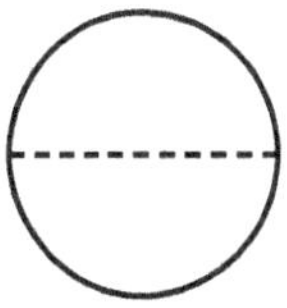

_____ equal parts

_____ halves

a **whole** pizza

a pizza cut in **half**

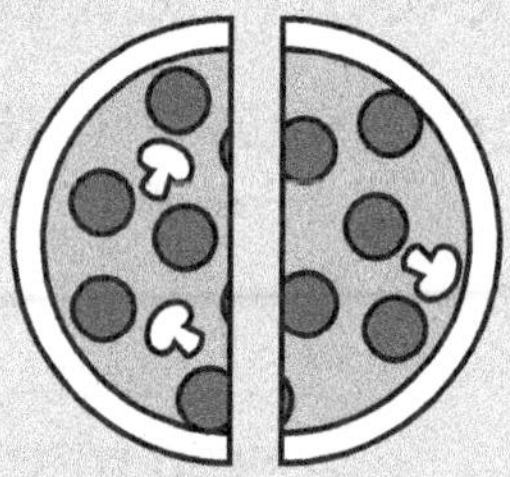

more than half → 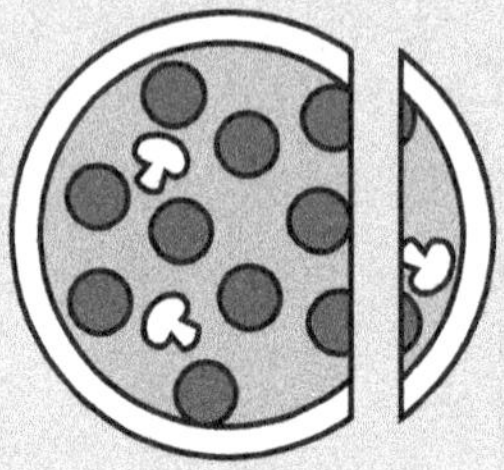 ← **less** than half

☐ Write **M** if the pizza part is more than half.

☐ Write **L** if the pizza part is less than half.

5. 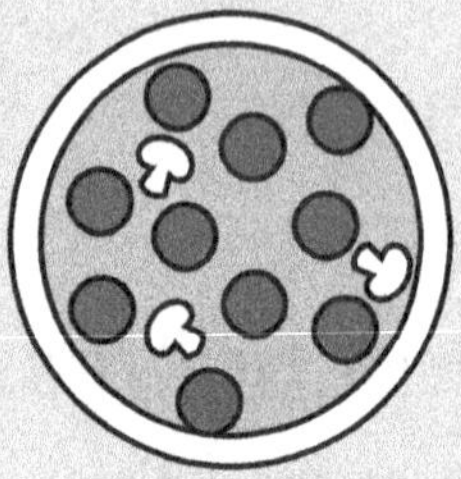

M

6. 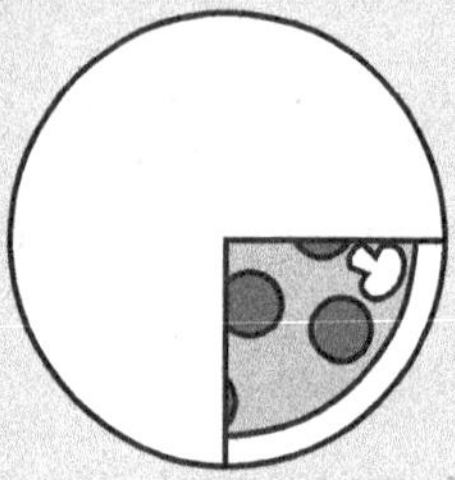

L

7. 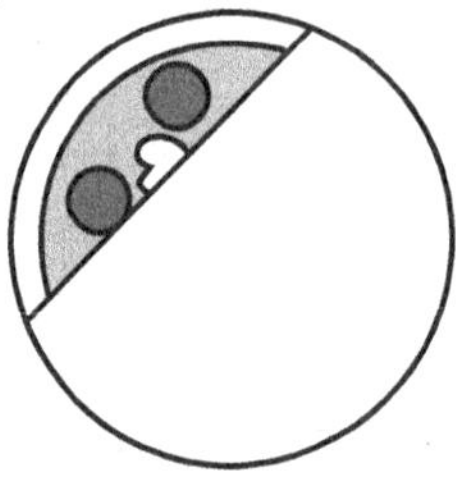

______

8. 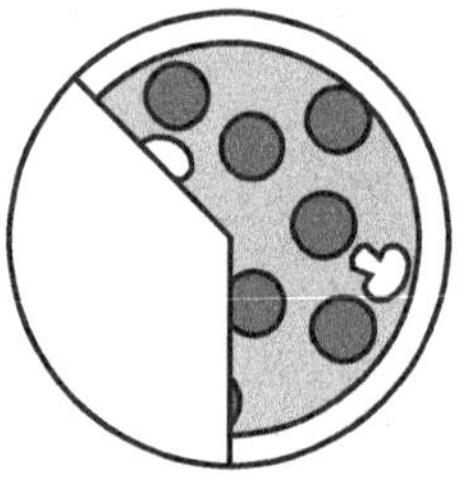

______

9. 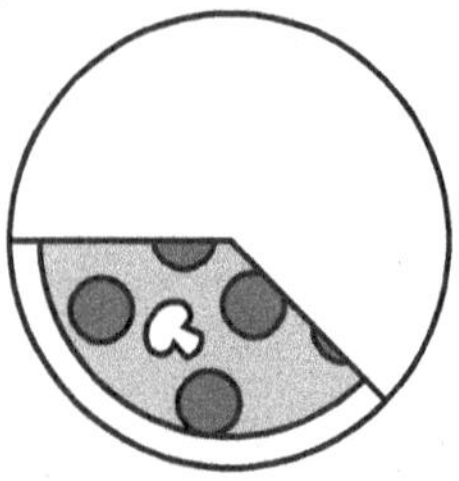

______

10. 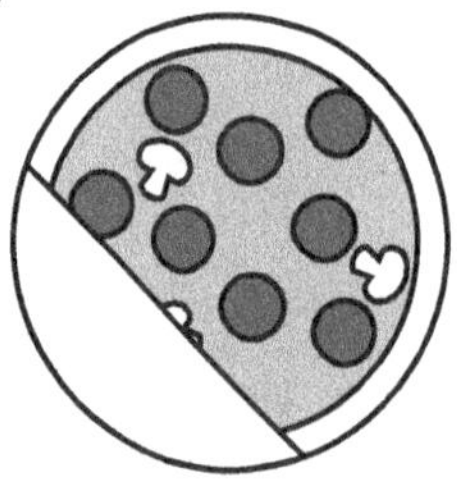

______

11. 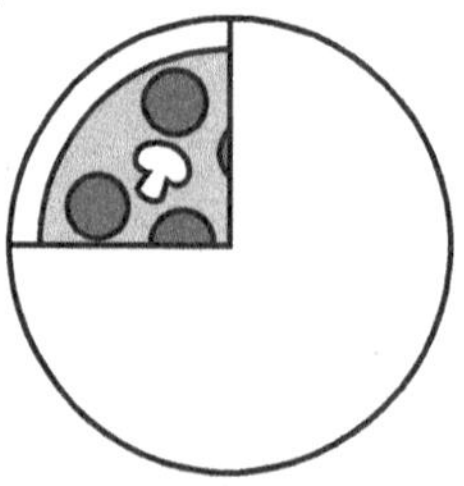

______

12. 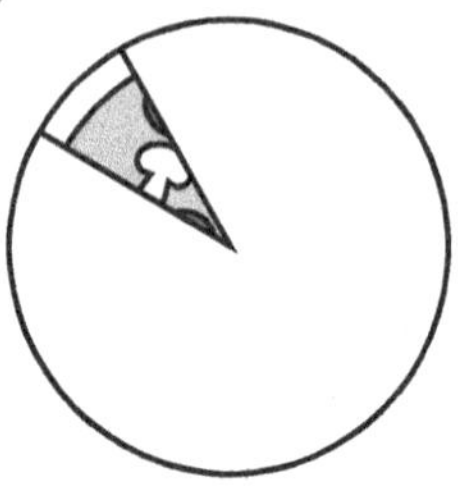

______

Here are 4 ways to fold a square in half.

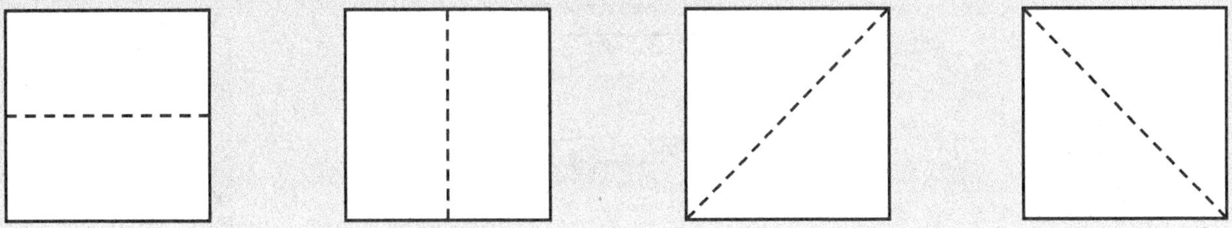

☐ Write **M** if more than half is shaded.

☐ Write **L** if less than half is shaded.

13.

M

14.

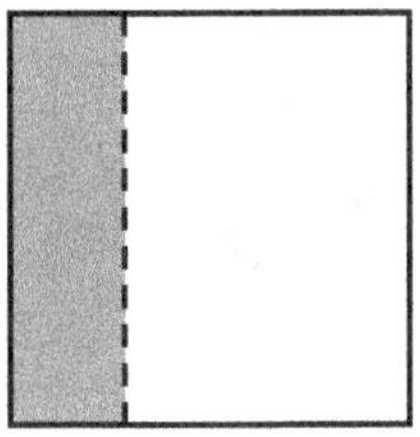

_____

15.

_____

16.

_____

17.

_____

18.

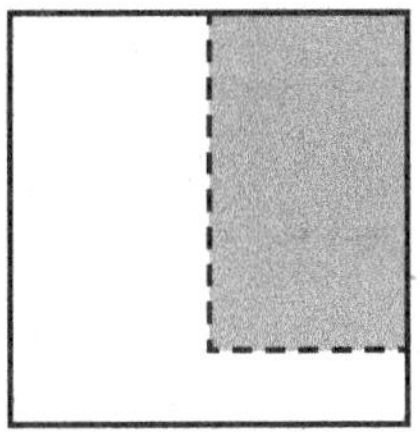

_____

19.

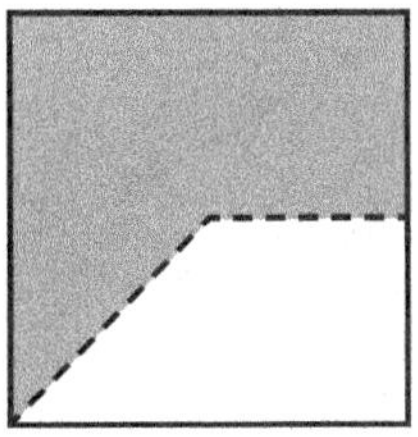

_____

20.

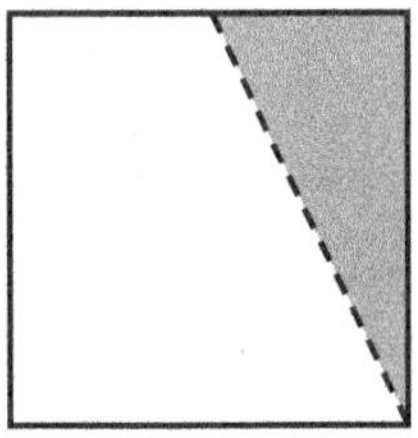

_____

21.

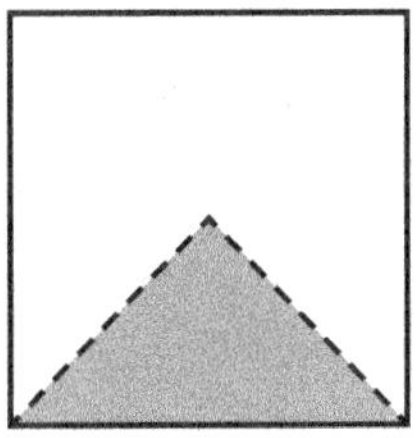

_____

☐ Circle the pictures that show a half.

22.

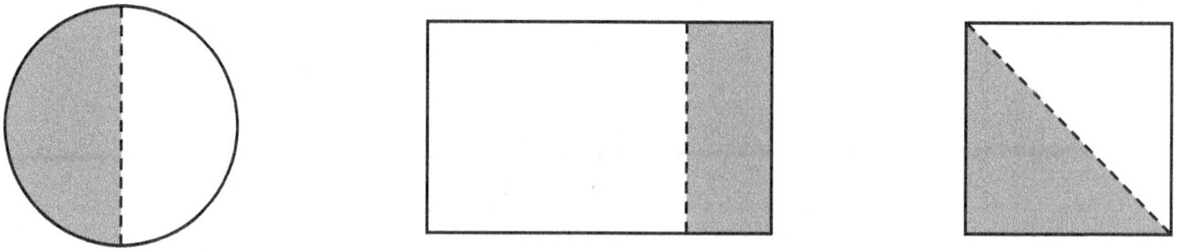

23.

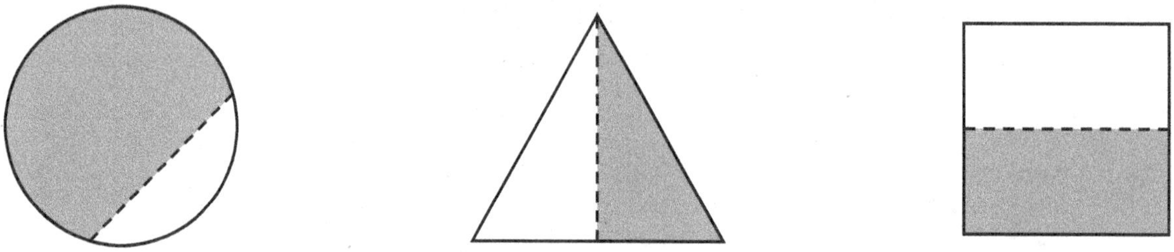

24.

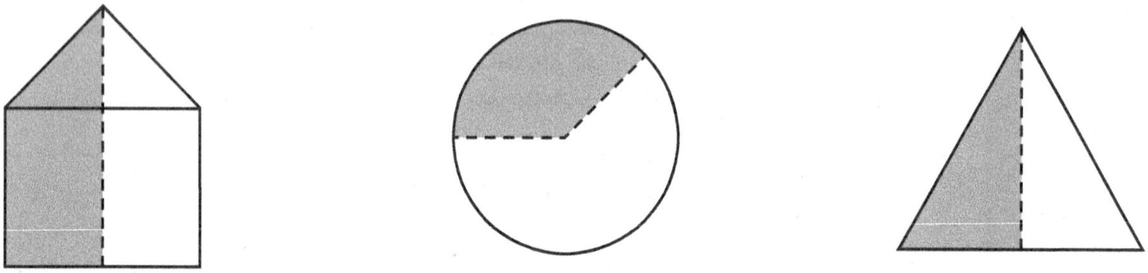

25.

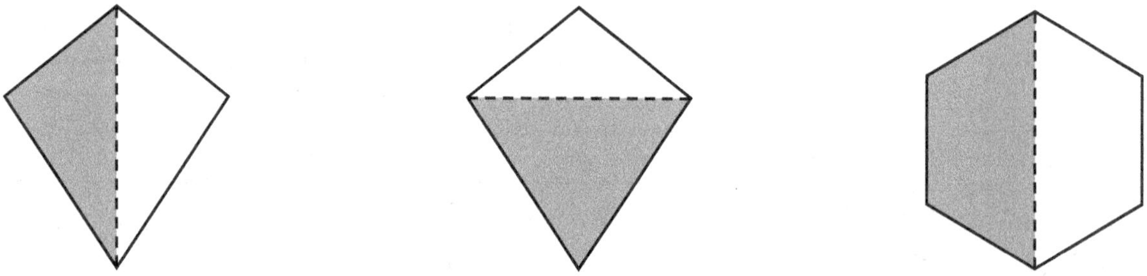

26. How many halves make a whole pizza? ______

GI-10 Quarter

The circle and the square are cut into 4 equal parts.

Each part is a **quarter** or a **fourth**.

4 fourths or 4 quarters

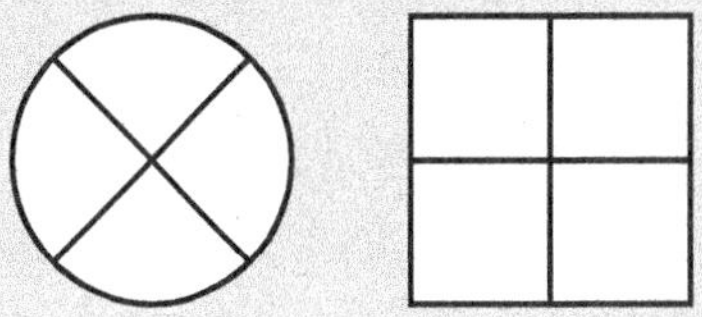

Is the circle cut into quarters?

☐ Write **yes** or **no**.

1.

yes

2.
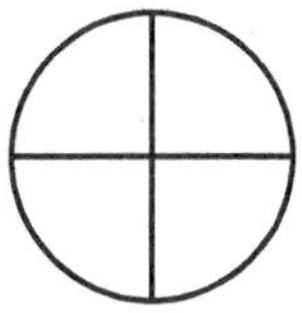
____

3.
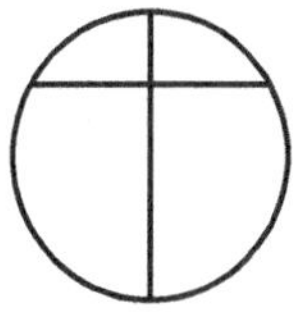
____

4.
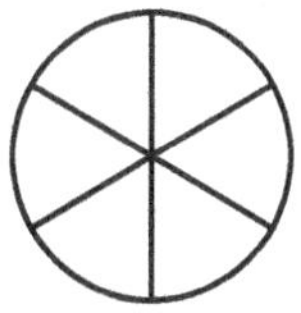
____

☐ Circle the pictures that have one quarter shaded.

5.

6.

7.

Here are 4 ways to fold a square into quarters.

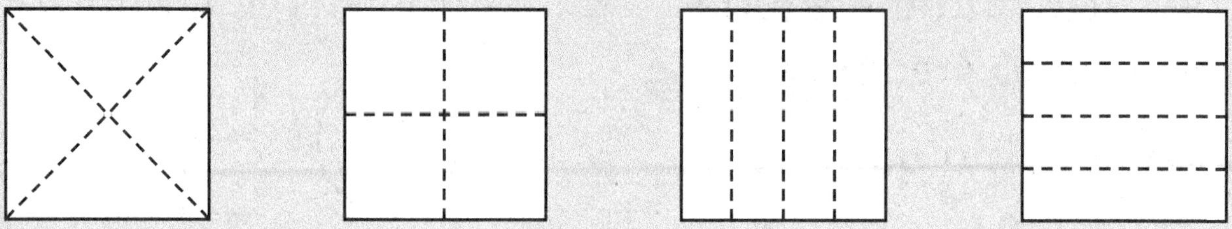

These are not quarters.

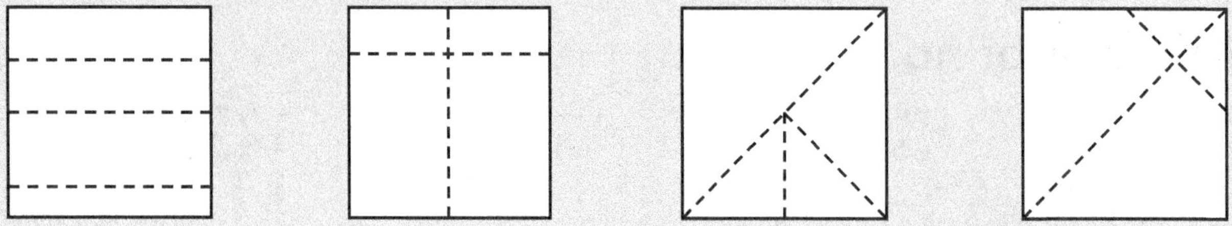

☐ Circle the pictures that have one quarter shaded.

8.

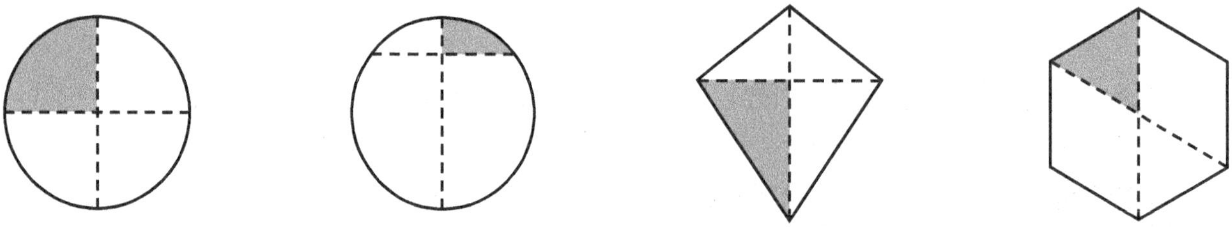

9.

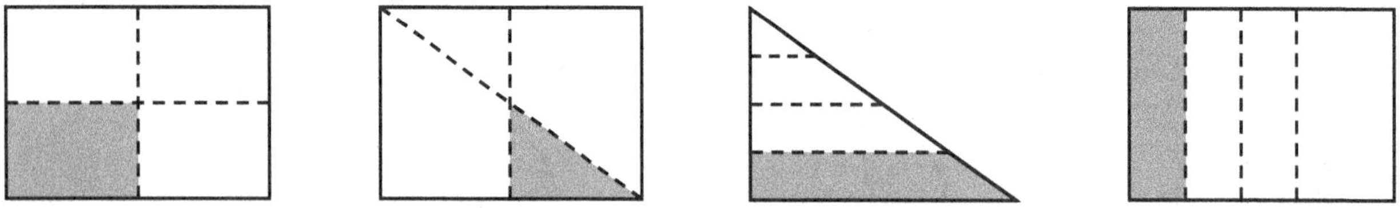

10.

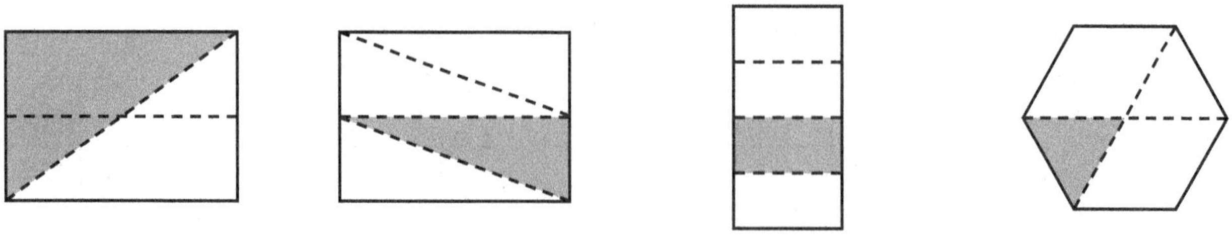

☐ Draw one more line to cut the shape into fourths.

☐ Check by counting.

11.

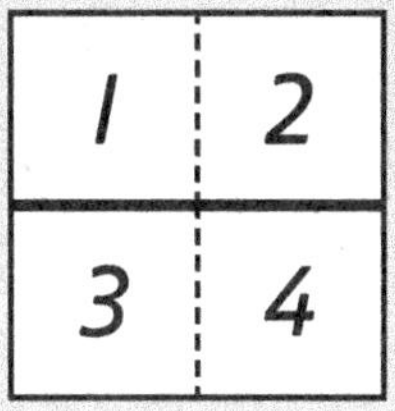

12.

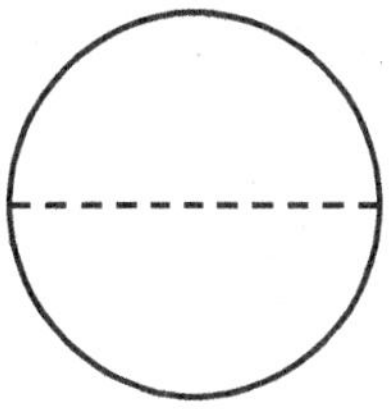

13.

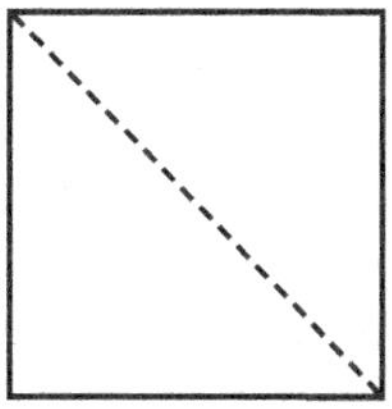

14.

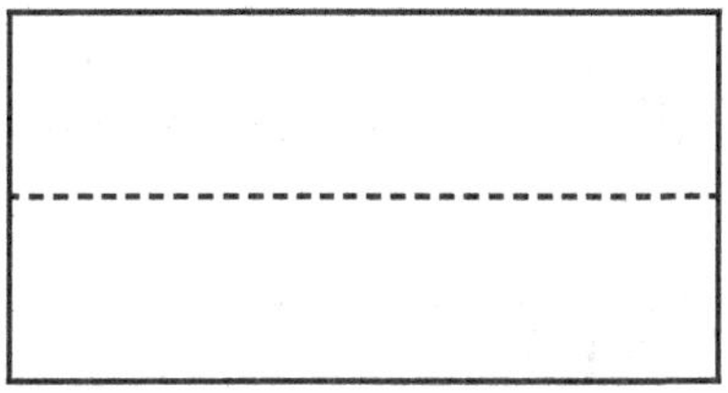

A whole shape has been cut into quarters.

☐ Draw the missing piece.

15.

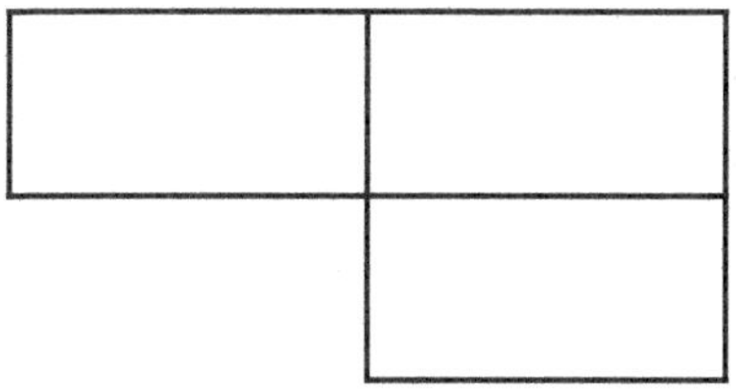

16.

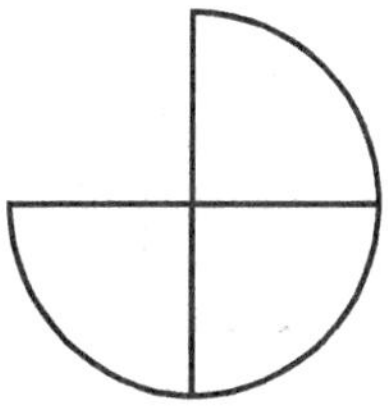

☐ Fill in the blanks.

17. How many quarters of a bagel do you need to make a whole bagel? ________________

18. Nina's bread is cut in fourths. Yu's bread is cut in halves. Who has smaller pieces? ________________

Fractions

☐ Write **half** or **fourth**.

1. There are 2 equal parts.

Each part is a

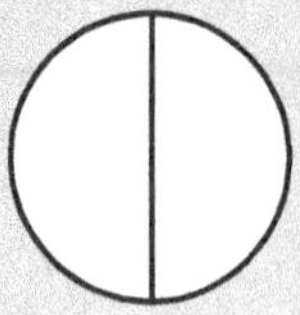

half ______.

2. There are 4 equal parts.

Each part is a

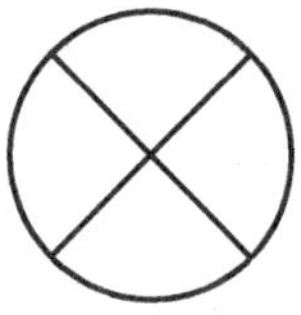

______.

3. There are 4 equal parts.

Each part is a

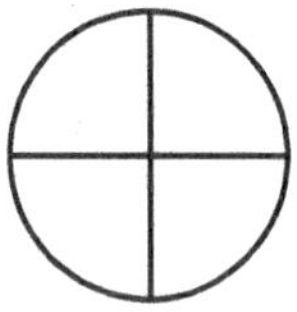

______.

4. BONUS There are ____ equal parts.

Each part is a

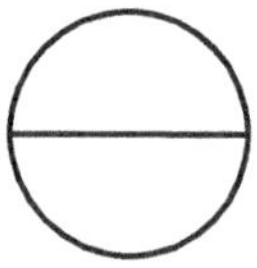

______.

☐ Fill in the blanks.

5.

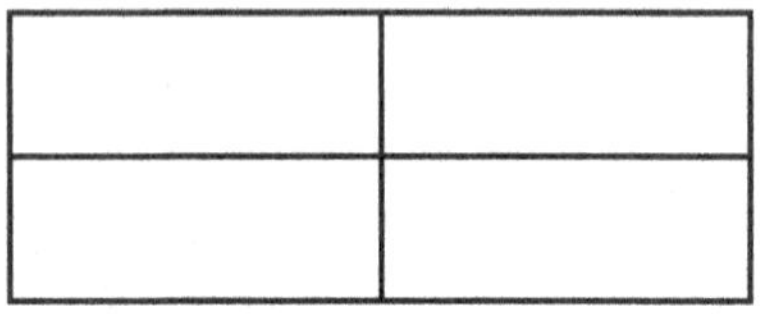

Each part is a ______
of a rectangle.

6.

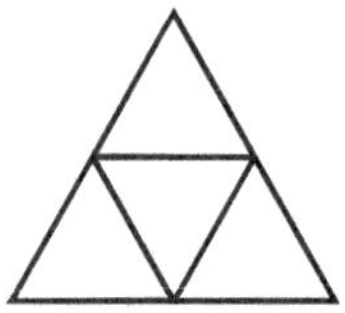

Each part is a ______
of a triangle.

7.

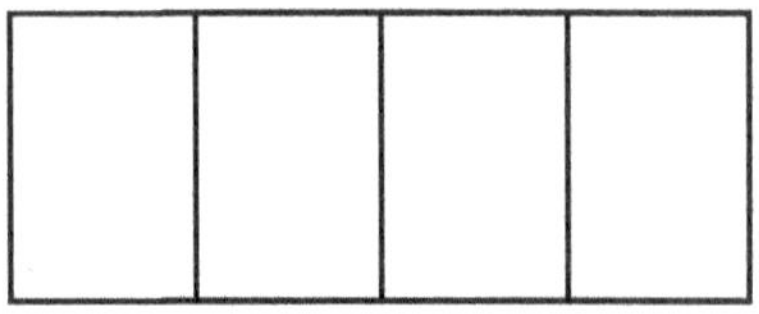

Each part is a ______

of a ______.

8.

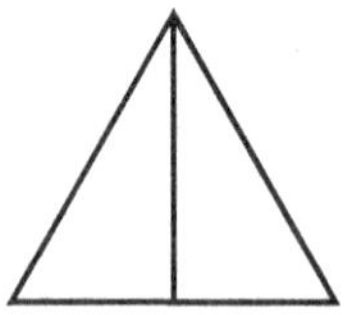

Each part is a ______

of a ______.

☐ Write how many parts are shaded.

9.

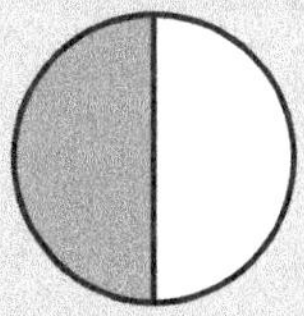

__1__ of __2__ parts.

10.

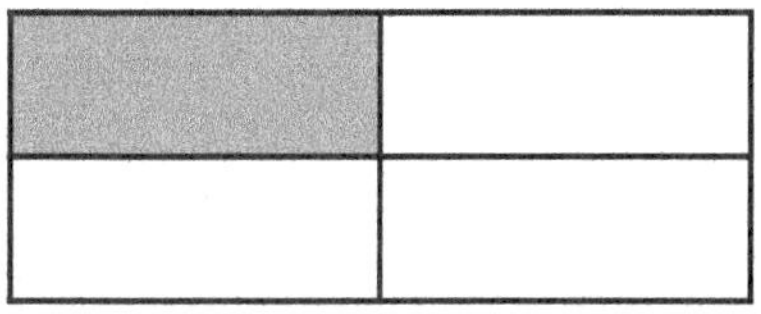

__1__ of ______ parts.

11.

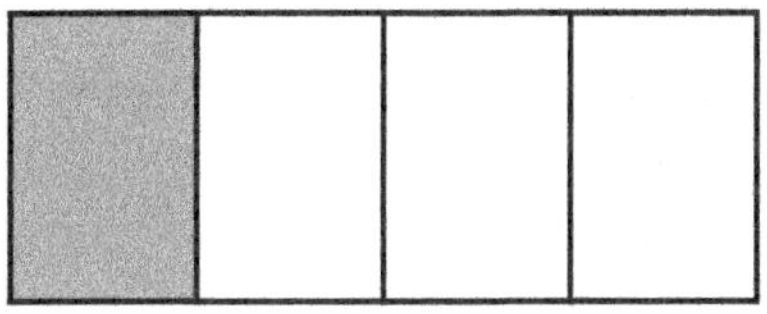

______ of ______ parts.

12.

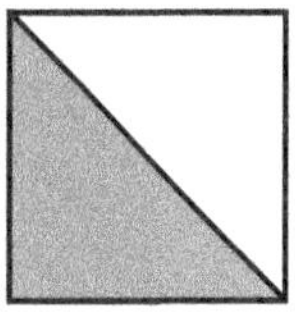

______ of ______ parts.

Half the shape is missing.

☐ Draw the missing half.

13.

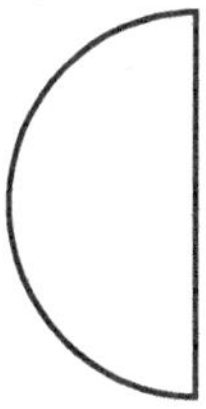

14.

☐ Circle the correct answer.

15. Mike wants to share a sandwich equally with his friend. What part should he give his friend?

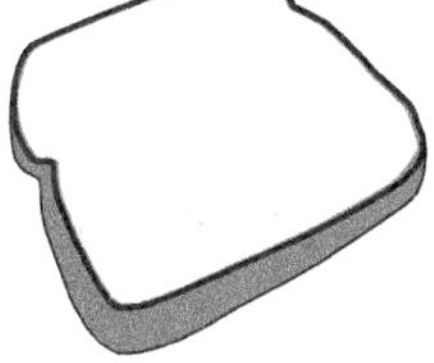

a half   a quarter

GI-12 Composing Shapes

yellow green red blue

☐ Color.

1.

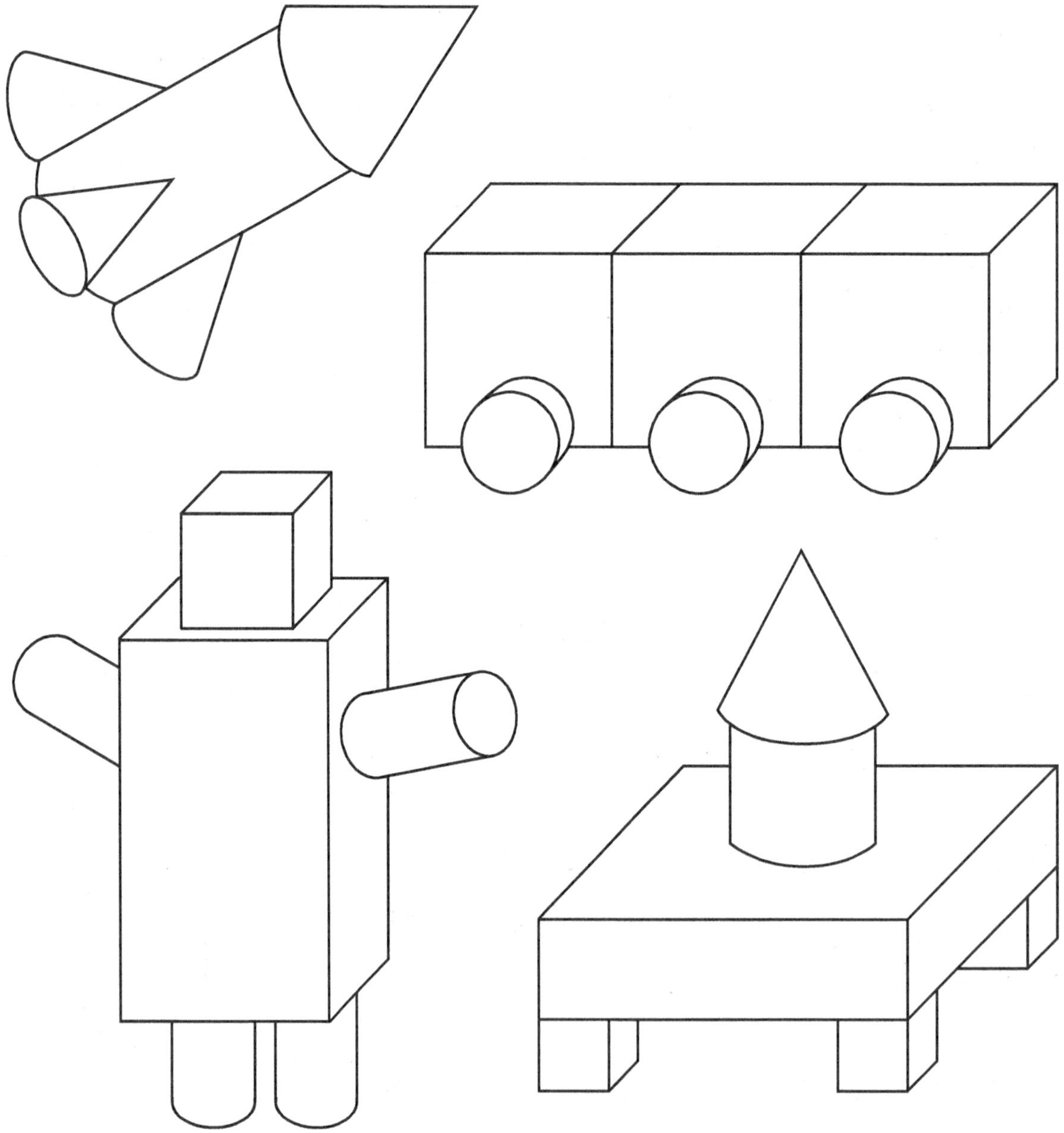

cones

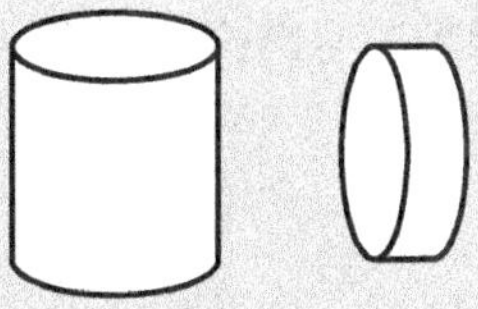

cylinders

☐ Color the cones blue.

☐ Color the cylinders red.

2.

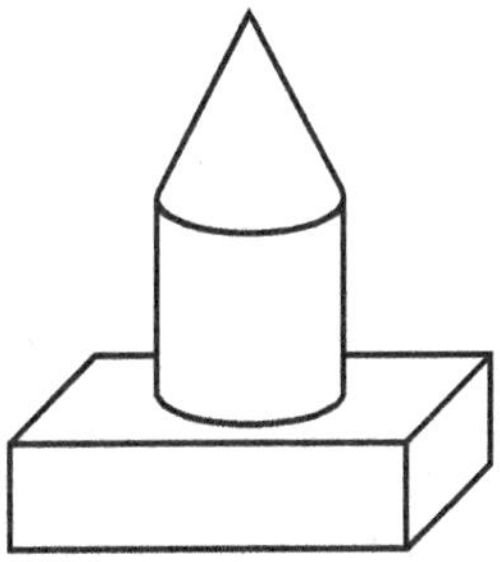

3.

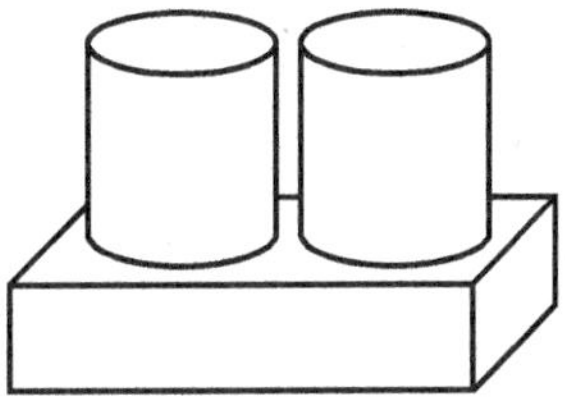

4.

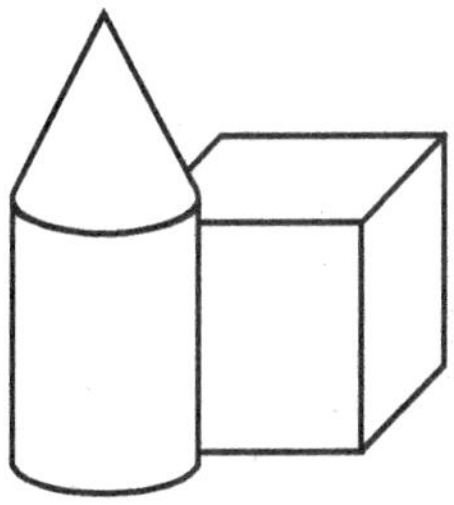

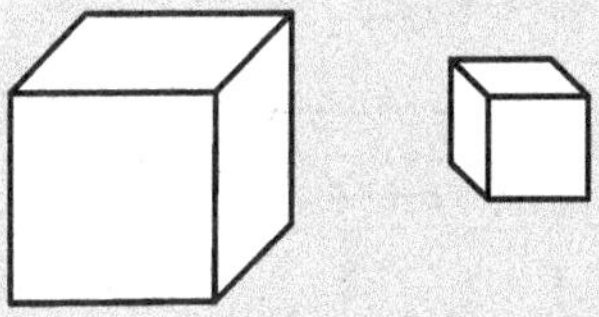

cubes

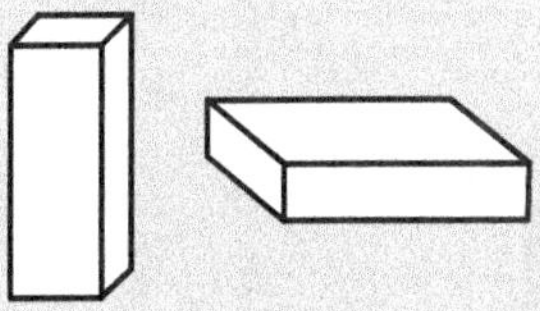

prisms

☐ Color the cubes yellow.

☐ Color the prisms green.

5.

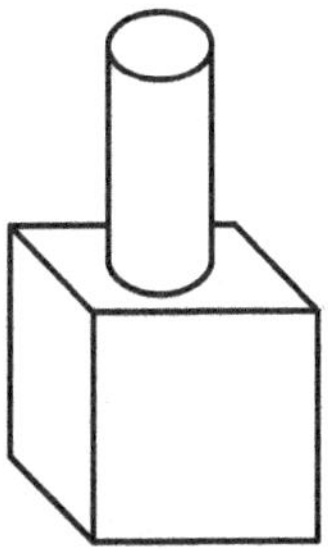

6.

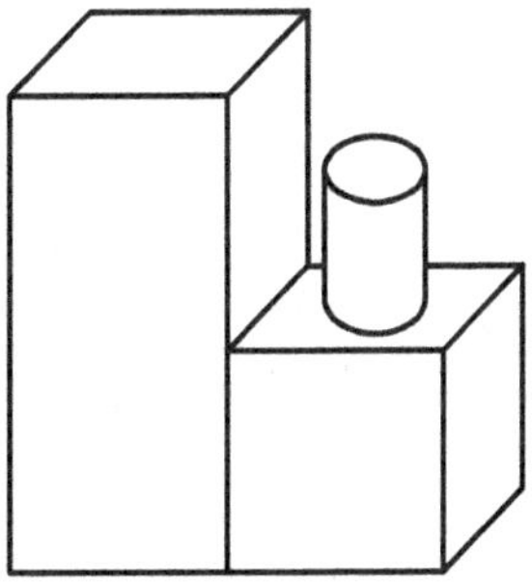

7.

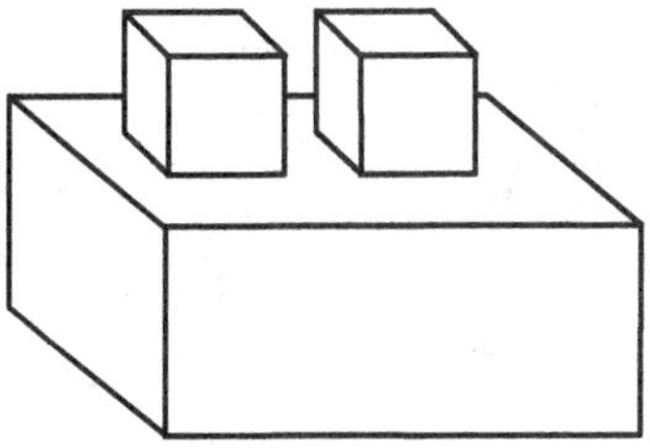

Mona makes a building from blocks.

☐ Circle the shapes she uses.

8.

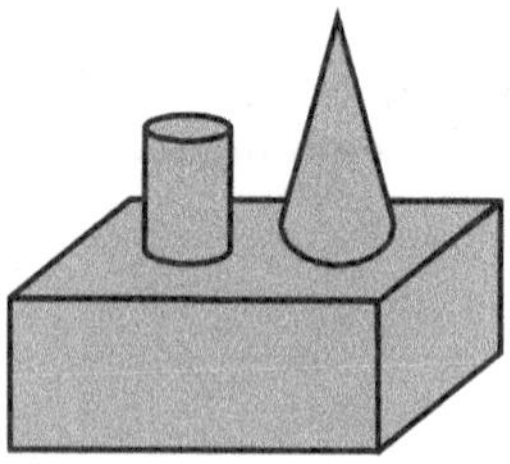

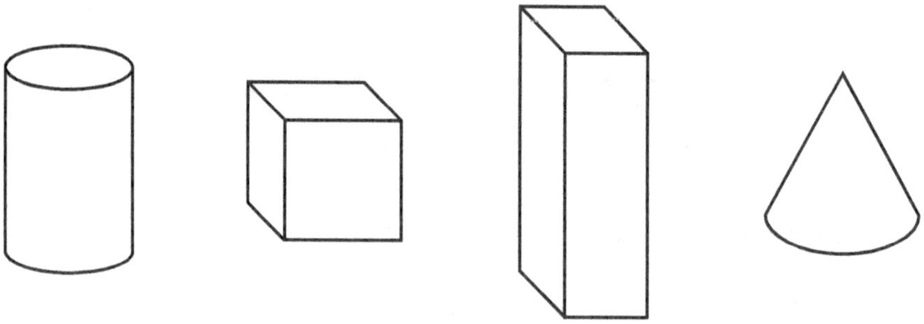

9.

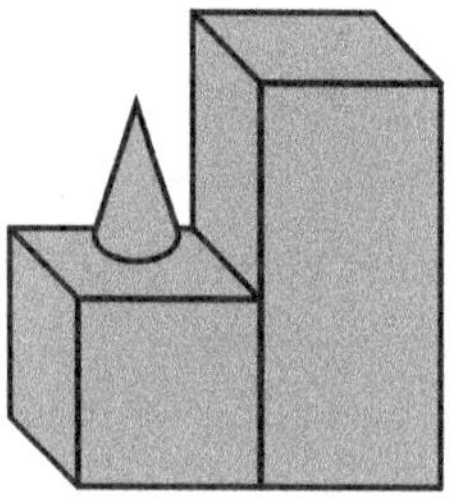

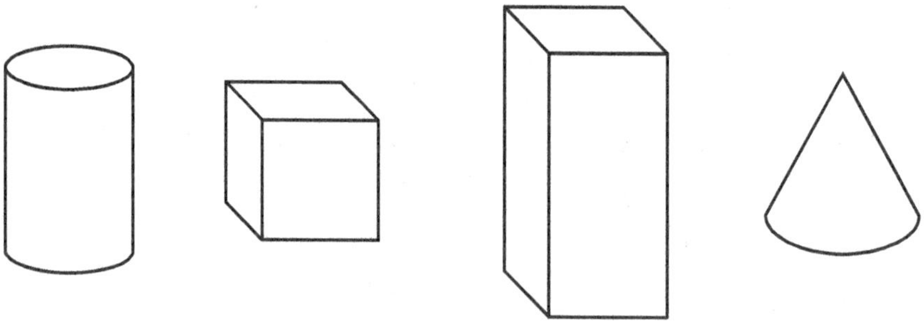

10.

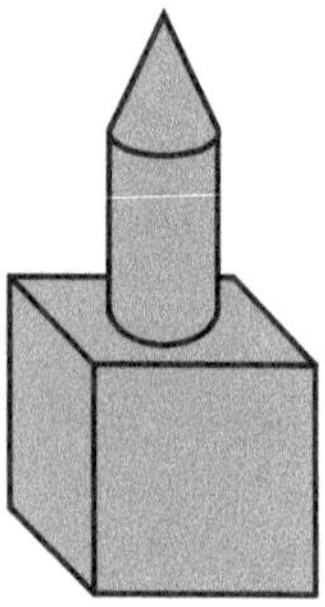

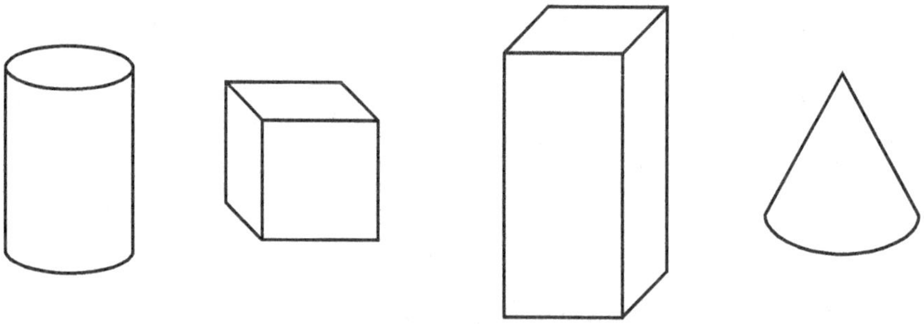

11. BONUS

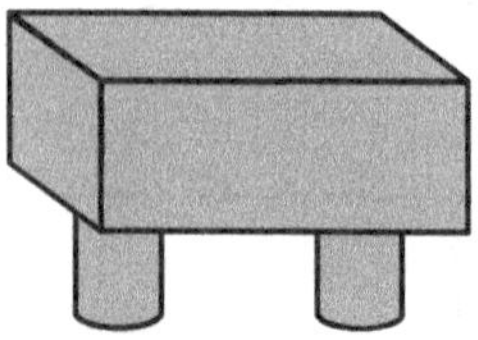

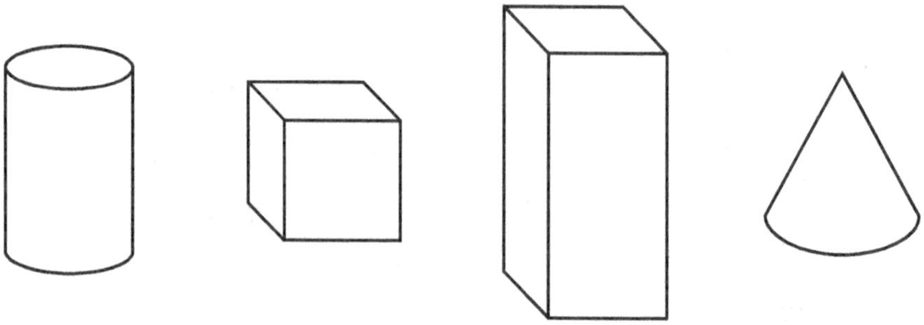

MDI-19 Sorting

You can **sort** anything into **groups**.
The things you sort are called **data**.
Some data may not belong in a group.

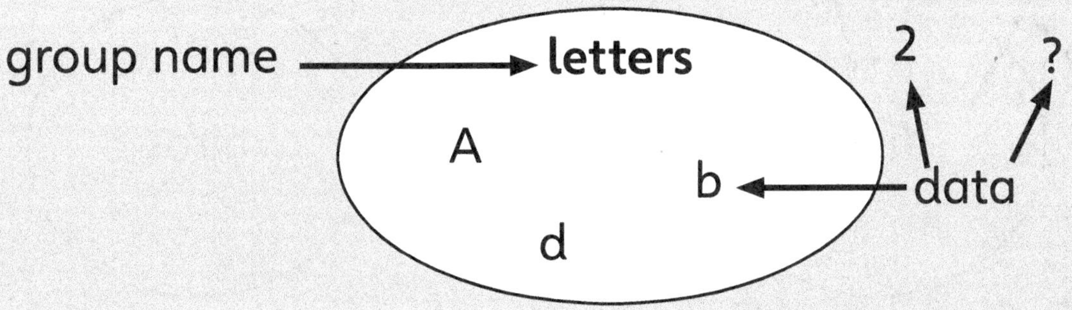

☐ Sort the data.

1. ~~t~~ ~~5~~ ! E = 7 s

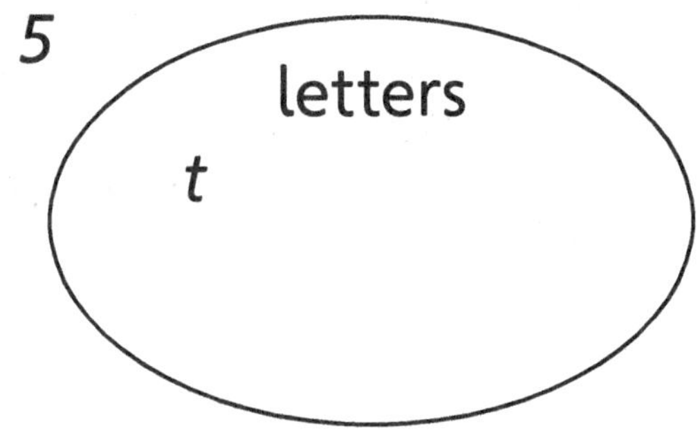

2. E 3 7 s < 8 d 9

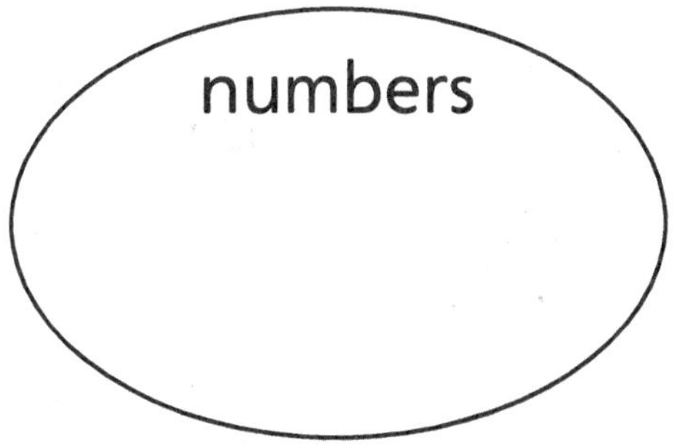

3. A b 3 △ 4 k 12 R

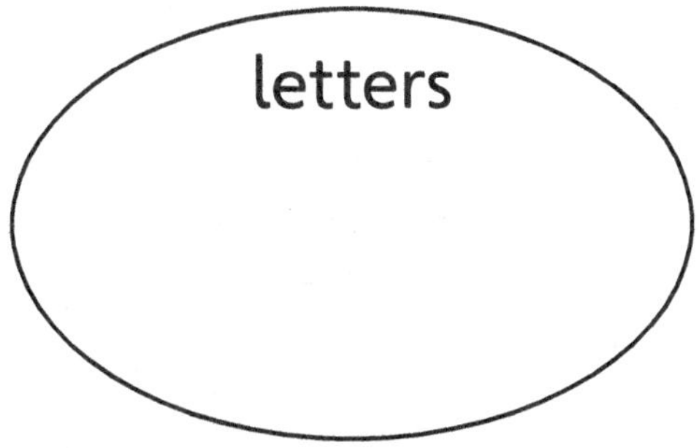

4. A b 3 △ 4 k 12 R

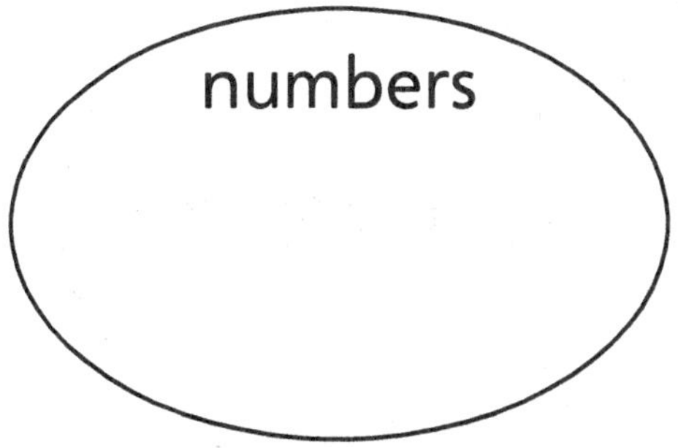

Sort the data.

5.

fish

6.

birds

7.

animals

8.

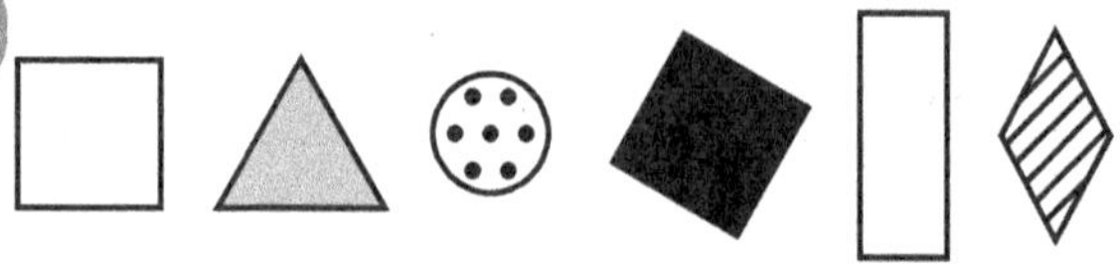

squares

9.

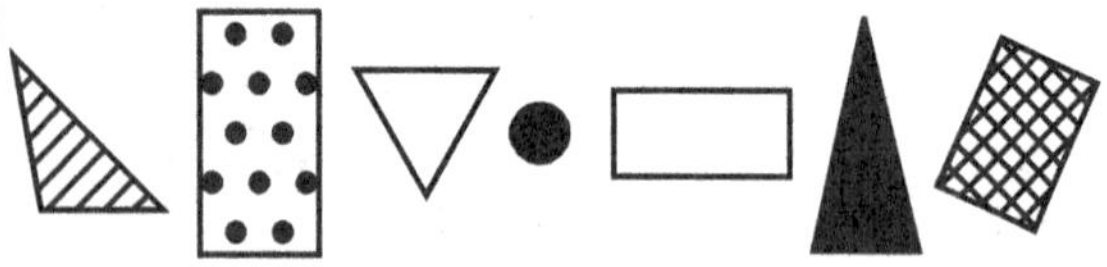

triangles

10.

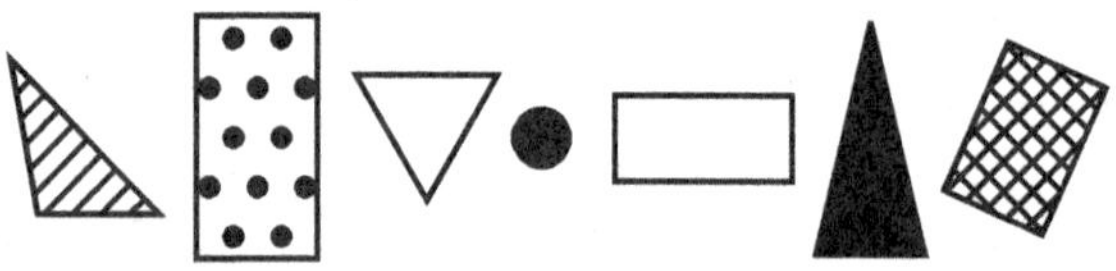

rectangles

MDI-20 Sorting into Many Groups

☐ Sort the data.

1.

fruits

vegetables

2.

shoes

boots

3.

animals with spots

animals with stripes

Some objects do not belong in any group.

☐ Sort the data.

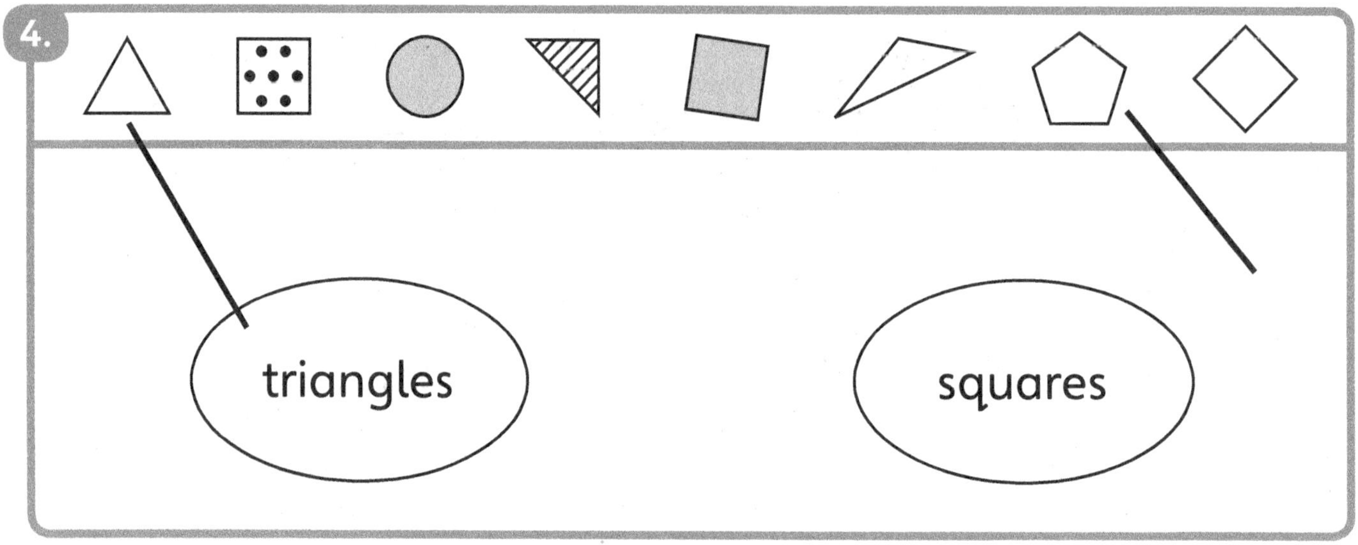

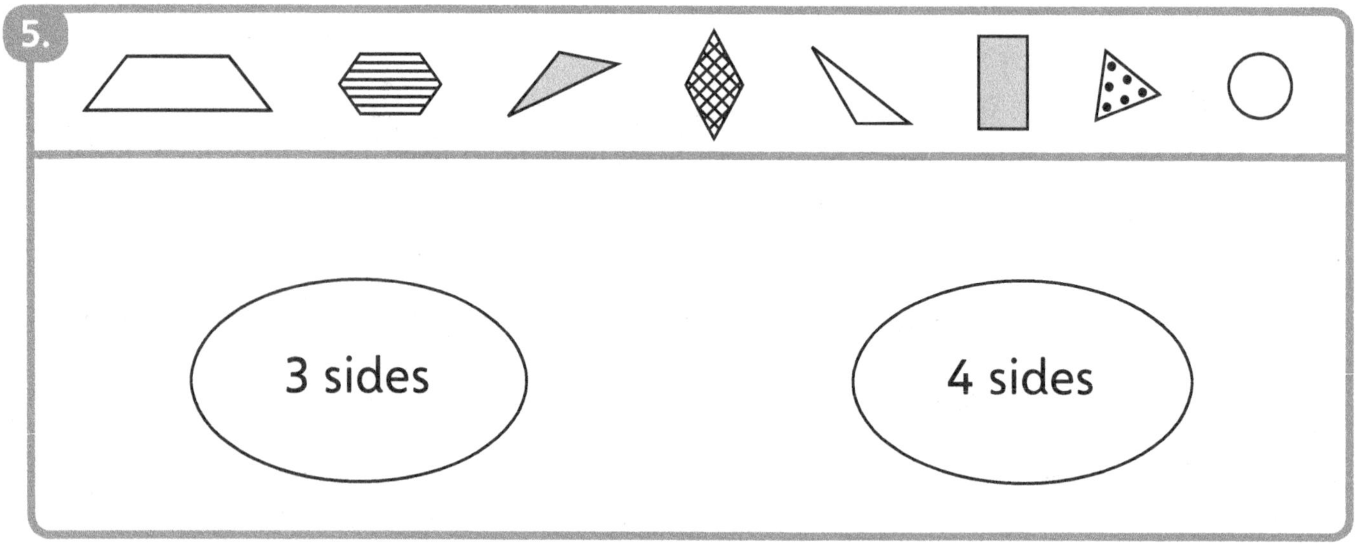

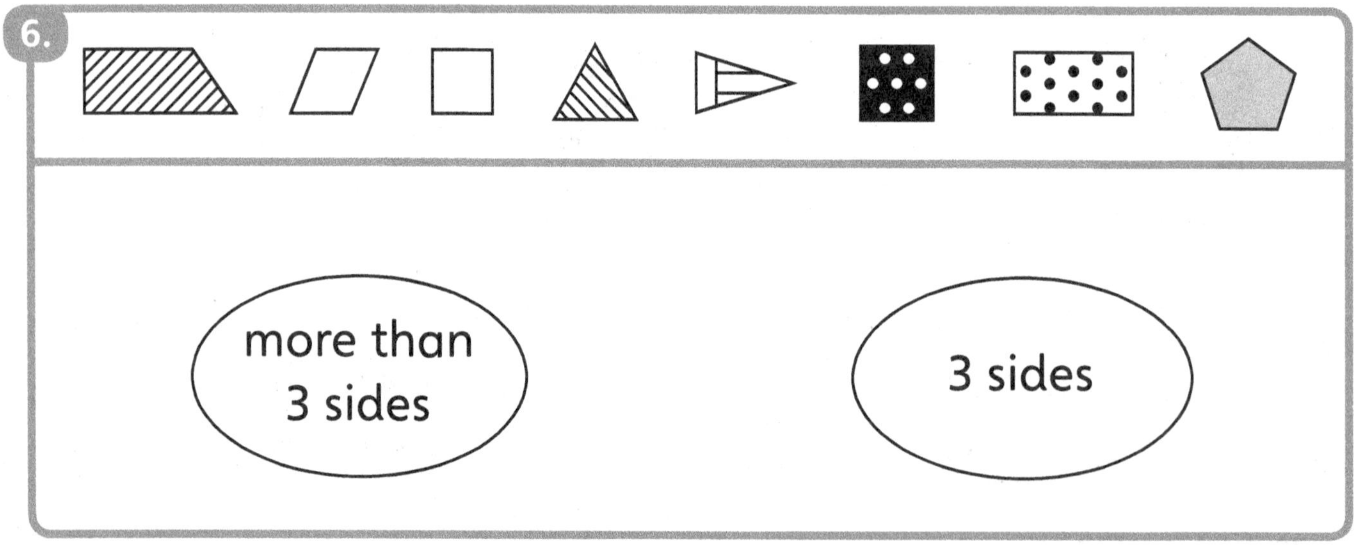

Sort the data.

7.

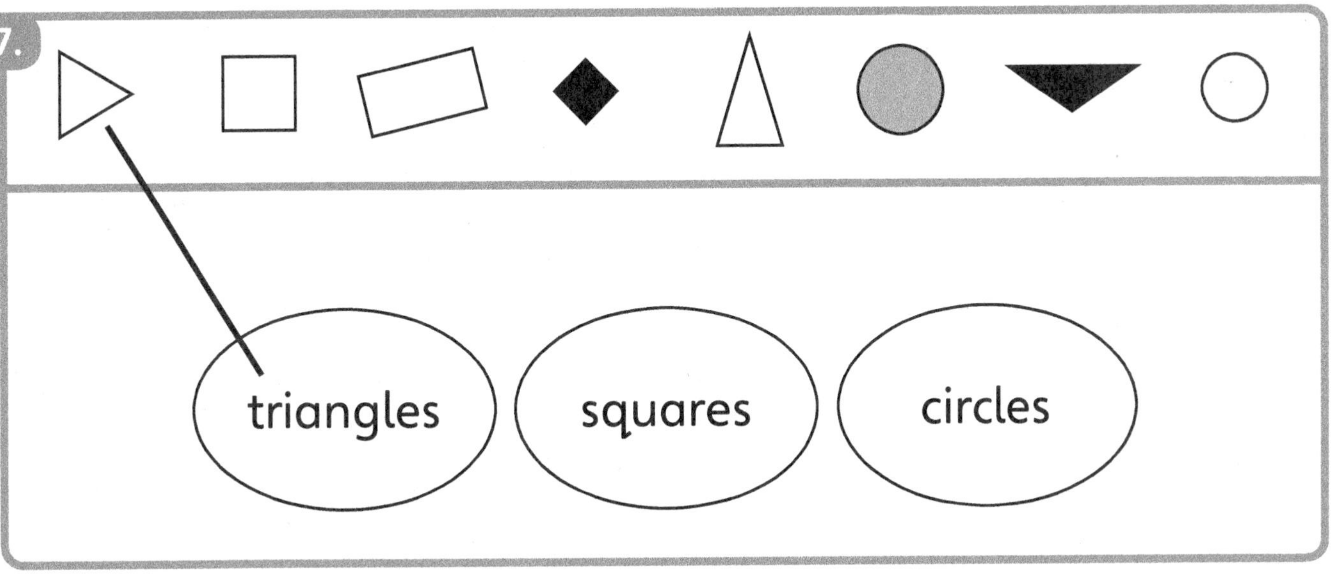

8.

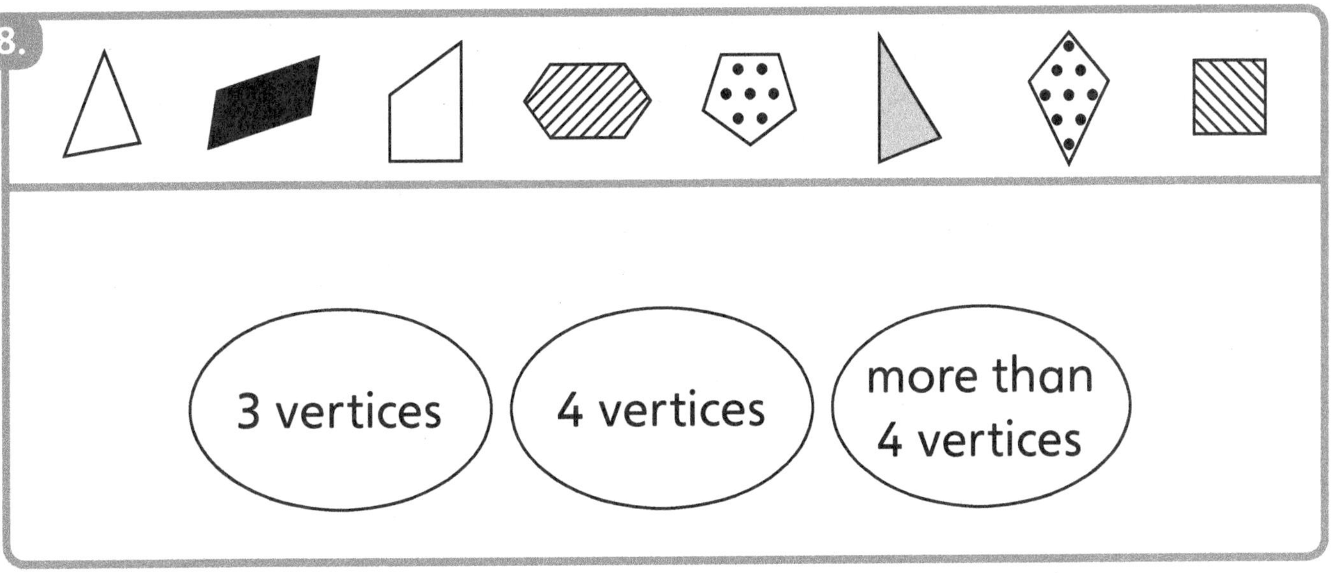

9.

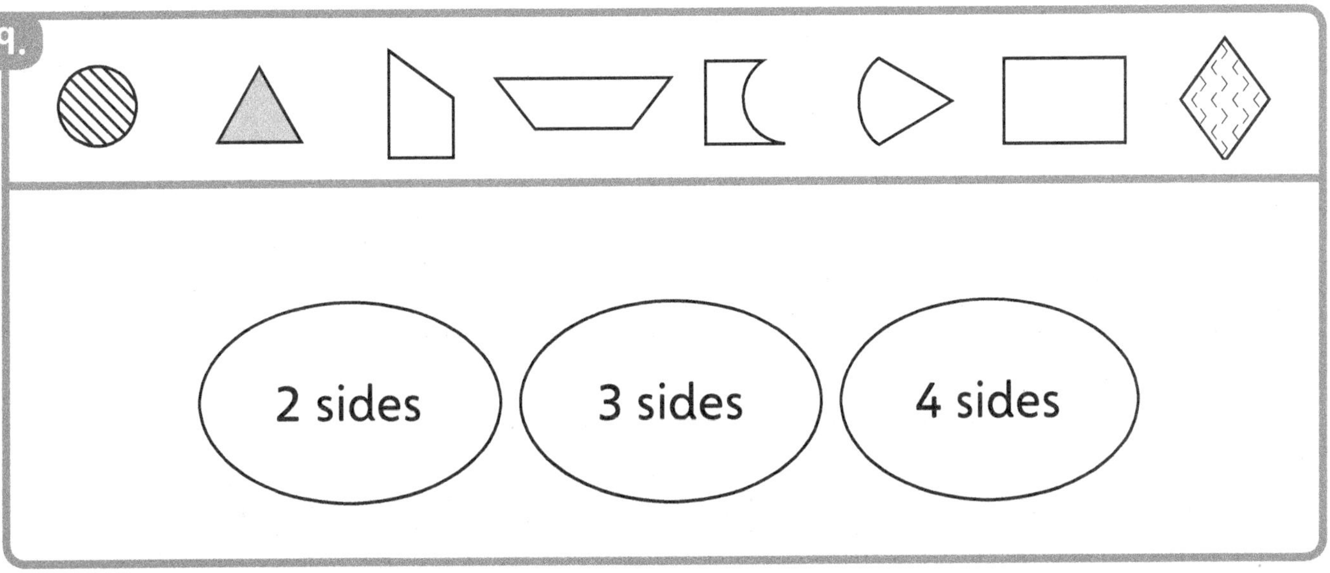

MDI-21 Sorting and Graphing

- [ ] Sort the data.
- [ ] Copy the sorted data into the rows.

1.

~~L~~ a g m 4 5 b E 8 N

L

letters

numbers

Letters	L						
Numbers							

2.

in red man no yes at of

3 letters

2 letters

3 letters				
2 letters				

☐ Sort the data.

☐ Copy the sorted data into the rows.

3.

it bed a on let too me

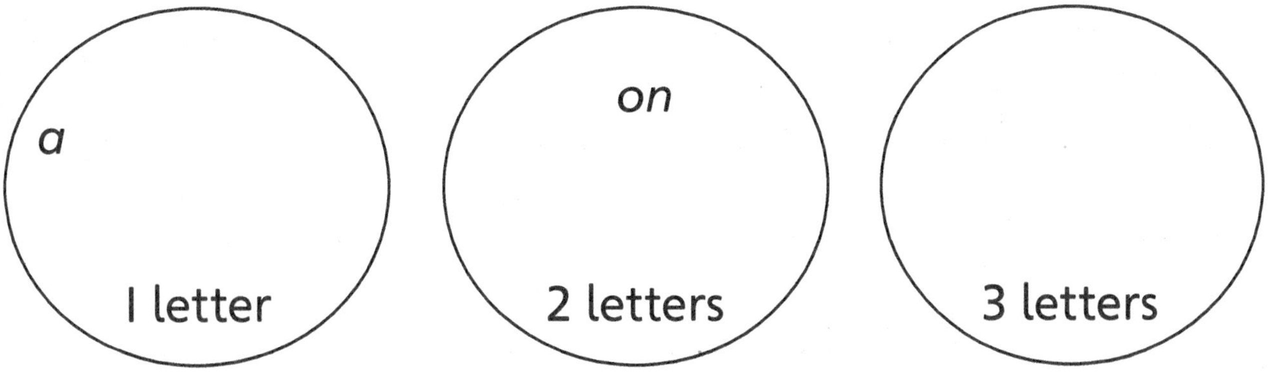

1 letter	*a*			
2 letters	*on*			
3 letters				

Look at your table.

☐ Fill in the blanks.

4. How many words with 2 letters did you sort? ______

5. How many words with 3 letters did you sort? ______

6. How many words did you sort altogether? ______

Picture Graphs

☐ Fill in the blanks.

1.

Lunch Time

__5__ at home

__3__ at school

More students eat lunch __at home__________.

2.

Mitts or Gloves

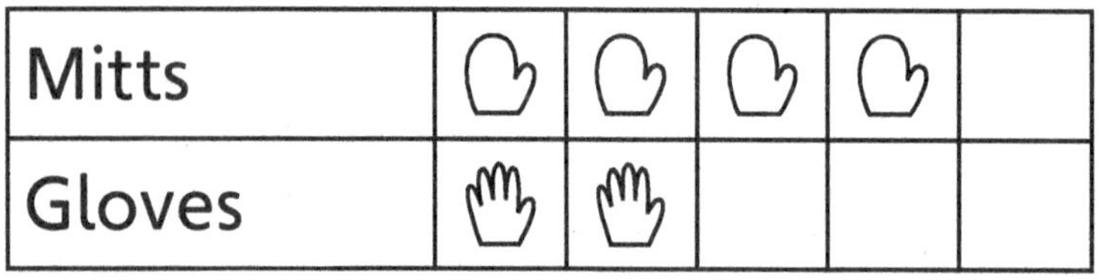

______ wear mitts

______ wear gloves

More students wear ____________________.

3.

Kate's Clothes

Kate has ______ skirts.

She has ______ shorts.

She has fewer __________

than ____________.

4.

Marco's Garden

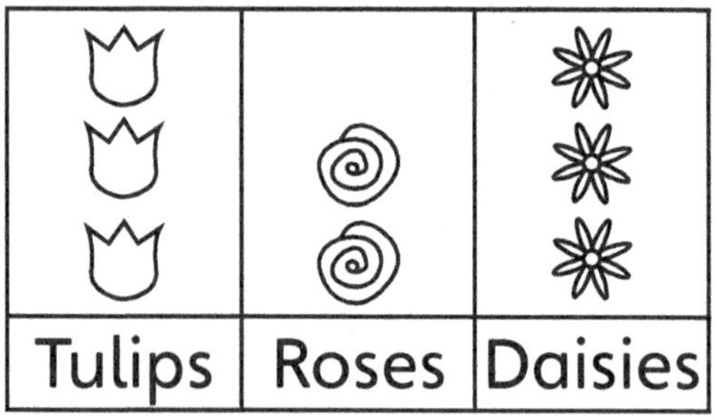

Marco has ______ roses.

He has the same

number of __________

and __________.

☐ How many more or fewer?

5.

Getting to School

Bus							
Walk							

[3] bus

[7] walk

[7] − [3] = 4

4 more students walk to school than take a bus.

6.

Drinks at Lunch

Juice							
Water							

☐ juices

☐ waters

☐ − ☐ = ____

______ more people drink juice than water.

7.

Birds We Saw

Jays							
Sparrows							

☐ jays

☐ sparrows

☐ − ☐ = ____

We saw ______ more sparrows than jays.

☐ Draw ○ to show the data.

☐ How many more or fewer?

8.

Sports We Play

Hockey	○	○	○		
Soccer					

3 play hockey

5 play soccer

_____ more students play _________ than _________.

9.

Hand We Write With

Left hand						
Right hand						

2 left hand

6 right hand

_____ fewer people write with their __________ hand.

10.

Pets We Own

Cats					
Hamsters					
Fish					

4 cats

5 hamsters

1 fish

We own 1 more *hamster* than __________.

We own 3 more ______________ than __________.

We own 4 more ______________ than __________.

MDI-23 Tally Charts

The **tally chart** shows a mark for each vote.

/ means 1 vote.

☐ Find the totals.

1.

Favorite Snack

	Tally	Total
Carrots	///	3
Apples	//	2
Muffins	////	4

2.

Favorite Pet

	Tally	Total
Dogs	//	
Cats	////	
Fish	//	

3.

Favorite Fruit

	Tally	Total
Kiwis	/	
Apples	////	
Grapes	///	

Which fruit is the most popular?

____

4.

Favorite Color

	Tally	Total
Red	/	
Yellow	//	
Blue	////	

Which color is the least popular?

____

☐ Write the totals.

5.

3 friends go hiking.
They each write down how many birds they see.

Number of Birds

	Tally	Total
Mike	///	3
Blanca	////	
Rani	/	

☐ Answer the questions.

6.

How many more birds does Mike see than Rani?

____

7.

Who sees the most birds?

____

8.

How many birds do Mike and Rani see altogether?
Show your work by writing a number sentence.

____

☐ Draw a picture graph from the tally chart.

9.

Favorite Vegetable

	Tally	Total
Peas	///	3
Corn	////	
Carrots	//	

Favorite Vegetable

Peas	○	○	○			
Corn						
Carrots						

○ means 1.

10.

Favorite Subject

	Tally	Total
Art	///	
Math	////	
Gym	///	

Favorite Subject

Art						
Math						
Gym						

○ means 1.

11. BONUS

𝍸 means 5.

Favorite Animal

	Tally	Total
Lions	𝍸 /	6
Whales	///	
Bears	𝍸 //	

Favorite Animal

Lions							
Whales							
Bears							

○ means 1.

MDI-24 More Charts and Picture Graphs

☐ Use the picture graph to answer the questions.

1.

Camp Classes

How many campers go to gym? ______

How many more campers go to music than art? ______

2. BONUS

Hair Color

Blond	✓	✓	✓	✓	✓		
Brown	✓	✓	✓	✓	✓	✓	✓
Red	✓	✓	✓				
Black	✓	✓	✓	✓	✓	✓	

How many fewer students have red hair than brown hair? ______

How many students have blond or black hair? ______

Fruit	Total
Apples	8
Bananas	4
Grapes	7

A teacher asks students to name their favorite fruit.

She writes the number of students who pick each fruit in a table.

☐ Make a picture graph of the data. Use ○ for one fruit.

3.

Apples								
Bananas								
Grapes								

☐ Use the picture graph to answer the question.

4. How many more students like apples than bananas?

5. How many fewer students like bananas than grapes?

6. How many students are in the class?

7. Ask two new questions about the data.